Lecture Notes in Mathematics 2185

Editors-in-Chief:
Jean-Michel Morel, Cachan
Bernard Teissier, Paris

Advisory Board:
Michel Brion, Grenoble
Camillo De Lellis, Zurich
Alessio Figalli, Zurich
Davar Khoshnevisan, Salt Lake City
Ioannis Kontoyiannis, Athens
Gábor Lugosi, Barcelona
Mark Podolskij, Aarhus
Sylvia Serfaty, New York
Anna Wienhard, Heidelberg

More information about this series at http://www.springer.com/series/304

Bernard Candelpergher

Ramanujan Summation
of Divergent Series

 Springer

Bernard Candelpergher
Laboratoire J.A. Dieudonné. CNRS
Université de Nice
Côte d'Azur, Nice, France

ISSN 0075-8434   ISSN 1617-9692   (electronic)
Lecture Notes in Mathematics
ISBN 978-3-319-63629-0   ISBN 978-3-319-63630-6   (eBook)
DOI 10.1007/978-3-319-63630-6

Library of Congress Control Number: 2017948388

Mathematics Subject Classification (2010): 40D05, 40G05, 40G10, 40G99, 30B40, 30B50, 11M35, 11M06

Printed on acid-free paper

This Springer imprint is published by Springer Nature
The registered company is Springer International Publishing AG
The registered company address is: Gewerbestrasse 11, 6330 Cham, Switzerland

Introduction: The Summation of Series

The strange sums

$$\sum_{n\geq 0}^{+\infty} n = 0 + 1 + 2 + 3 + 4 + 5 + 6 + 7 + 8 + 9 + \dots$$

and

$$\sum_{n\geq 0}^{+\infty} n^3 = 0 + 1^3 + 2^3 + 3^3 + 4^3 + 5^3 + 6^3 + 7^3 \dots$$

appear in physics about the study of the Casimir effect which is the existence of an attractive force between two parallel conducting plates in the vacuum.

These series are examples of *divergent series* in contrast to convergent series; the notion of convergence for a series was introduced by Cauchy in his *Cours d'Analyse* in order to avoid frequent mistakes in working with series. Given a series of complex numbers $\sum_{n\geq 0} a_n$, Cauchy considers the sequence of the partial sums

$$s_0 = 0$$

$$s_1 = a_0$$

$$s_2 = a_0 + a_1$$

$$\dots$$

$$s_n = a_0 + \dots + a_{n-1}$$

and says that *the series* $\sum_{n\geq 0} a_n$ *is convergent* if and only if the sequence (s_n) has a finite limit when n goes to infinity. In this case the *sum of the series* is defined by

$$\sum_{n=0}^{+\infty} a_n = \lim_{n\to+\infty} s_n$$

The classical Riemann series $\sum_{n\geq 1} \frac{1}{n^s}$ is convergent for every complex number s such that $\mathrm{Re}(s) > 1$ and defines the *Riemann zeta function* defined for $\mathrm{Re}(s) > 1$ by $\zeta : s \mapsto \sum_{n=1}^{+\infty} \frac{1}{n^s}$.

Non-convergent series are *divergent series*. For $\mathrm{Re}(s) \leq 1$ the Riemann series is a divergent series and does not give a finite value for the sums that appear in the Casimir effect. A possible strategy to assign a finite value to these sums is to perform an analytic continuation of the zeta function this has been done by Riemann (Edwards 2001) who found an integral formula for $\zeta(s)$ which is valid not only for $\mathrm{Re}(s) > 1$ but also for $s \in \mathbb{C}\backslash\{1\}$. By this method we can assign to the series $\sum_{n\geq 1} n^k$ with $k > -1$ the value $\zeta(-k)$; we get, for example,

$$\sum_{n\geq 1} n^0 = 1 + 1 + 1 + 1 + 1 + 1 + \ldots \mapsto \zeta(0) = -\frac{1}{2}$$

$$\sum_{n\geq 1} n^1 = 1 + 2 + 3 + 4 + 5 + 6 + \ldots \mapsto \zeta(-1) = -\frac{1}{12}$$

$$\sum_{n\geq 1} n^2 = 1 + 2^2 + 3^2 + 4^2 + 5^2 + \ldots \mapsto \zeta(-2) = 0$$

$$\sum_{n\geq 1} n^3 = 1 + 2^3 + 3^3 + 4^3 + 5^3 + \ldots \mapsto \zeta(-3) = \frac{1}{120}$$

$$\ldots$$

For $k = -1$ we have the "harmonic series"

$$\sum_{n\geq 1} \frac{1}{n} = 1 + \frac{1}{2} + \frac{1}{3} + \frac{1}{4} + \frac{1}{5} + \frac{1}{6} + \frac{1}{7} + \frac{1}{8} + \frac{1}{9} + \ldots$$

which is easily proved to be a divergent series since the partial sums s_n verify

$$s_n = 1 + \frac{1}{2} + \ldots + \frac{1}{n} \geq \int_1^2 \frac{1}{x} dx + \int_2^3 \frac{1}{x} dx + \ldots + \int_n^{n+1} \frac{1}{x} dx = Log(n+1)$$

But the strategy of analytic continuation of the zeta function does not work in this case since ζ has a pole at $s = 1$ that is $\lim_{s\to 1} \zeta(s) = \infty$.

Divergent series appear elsewhere in analysis and are difficult to handle; for example, by using the preceding values of $\sum_{n\geq 1} n^k$, it seems that

$$-\frac{1}{12} - \frac{1}{2} = 1 + 2 + 3 + 4 + 5 + 6 + \ldots + (1 + 1 + 1 + 1 + 1 + 1 + \ldots)$$

$$= 2 + 3 + 4 + 5 + 6 + 7 + \ldots$$

$$= (1 + 2 + 3 + 4 + 5 + 6 + 7 + \ldots) - 1$$

$$= -\frac{1}{12} - 1$$

This absurdity shows that with divergent series we cannot use the classical rules of calculation, and for a given class of series, we need to define precisely some method of summation and its rules of calculation.

Before and after Cauchy, some methods of summation of series have been introduced by several mathematicians such as Cesaro, Euler, Abel, Borel, and others. These methods of summation assign to a series of complex numbers $\sum_{n\geq 0} a_n$ a number obtained by taking the limit of some means of the partial sums s_n. For example, the Cesaro summation assigns to a series $\sum_{n\geq 0}^{n} a_n$ the number

$$\overset{\mathcal{C}}{\sum_{n\geq 0}} a_n = \lim_{n\to+\infty} \frac{s_1 + \ldots + s_n}{n} \quad \text{(when this limit is finite)}$$

For the Abel summation, we take

$$\overset{\mathcal{A}}{\sum_{n\geq 0}} a_n = \lim_{t\to 1-} (1 - t) \sum_{n=0}^{+\infty} s_{n+1} t^n \quad \text{(when this limit is finite)}$$

where the series $\sum_{n\geq 0} s_{n+1} t^n$ is supposed to be convergent for every $t \in [0, 1]$. Note that this expression can be simplified since

$$(1 - t) \sum_{n=0}^{+\infty} s_{n+1} t^n = s_1 + \sum_{n=1}^{+\infty} (s_{n+1} - s_n) t^n = \sum_{n=0}^{+\infty} a_n t^n,$$

and we have

$$\overset{\mathcal{A}}{\sum_{n\geq 0}} a_n = \lim_{t\to 1-} \sum_{n=0}^{+\infty} a_n t^n \quad \text{(when this limit is finite)} ;$$

this gives, for example,

$$\overset{\mathcal{A}}{\sum_{n\geq 0}} (-1)^n = \lim_{t\to 1-} \sum_{n=0}^{+\infty} (-1)^n t^n = \lim_{t\to 1-} \frac{1}{1 + t} = \frac{1}{2}$$

The classical methods of summation use these types of means of partial sum and can be briefly presented in the following form.

Let T be a topological space of parameters and l some "limit point" of the compactification of T (if $T = \mathbb{N}$, then $l = +\infty$; if $T = [0, 1]$, then $l = 1$). Let $(p_n(t))_{n \in \mathbb{N}}$ be a family of complex sequences indexed by $t \in T$ such that for all $t \in T$ the series $\sum_{n \geq 0} p_n(t)$ is convergent; then we set

$$\sum_{n \geq 0}^{T} a_n = \lim_{t \to l} \frac{\sum_{n=0}^{+\infty} p_n(t) s_n}{\sum_{n=0}^{+\infty} p_n(t)}$$

when this limit is finite, and in this case, we say that the series $\sum_{n \geq 0} a_n$ is T-summable.

A theorem of Toeplitz (Hardy 1949) gives necessary and sufficient conditions on the family $(p_n(t))_{n \in \mathbb{N}}$ to ensure that in case of convergence this summation coincides with the usual Cauchy summation.

These summation methods verify the linearity conditions

$$\sum_{n \geq 0}^{T} (a_n + b_n) = \sum_{n \geq 0}^{T} a_n + \sum_{n \geq 0}^{T} b_n$$

$$\sum_{n \geq 0}^{T} C a_n = C \sum_{n \geq 0}^{T} a_n \text{ for every constant } C \in \mathbb{C}$$

and the usual *translation property*

$$\sum_{n \geq 0}^{T} a_n = a_0 + \sum_{n \geq 0}^{T} a_{n+1}$$

This last property which seems natural is in fact very restrictive. For example, the series $\sum_{n \geq 1} 1$ can't be T-summable since the translation property gives an absurd relation

$$\sum_{n \geq 0}^{T} 1 = 1 + \sum_{n \geq 0}^{T} 1$$

Thus, if we need a method of summation such that the sums $\sum_{n \geq 0}^{T} n^k$ are well defined for any integer k, then we *must abandon the translation property requirement* and find a way to define summation procedures other than the way of the "limit of means of partial sums."

This can be done by using a sort of generating function for the terms of the series. It is based on the following algebraic framework (Candelpergher 1995). Let E be a $\mathbb{C}$-vector space (in general a space of functions) equipped with a linear operator D

and a linear map $v_0 : E \to \mathbb{C}$. Given a sequence of complex numbers $(a_n)_{n \geq 0}$, we call an element $f \in E$ a generator of this sequence if

$$a_n = v_0(D^n f)$$

We can write formally

$$\sum_{n \geq 0} a_n = \sum_{n \geq 0} v_0(D^n f) = v_0 \left(\sum_{n \geq 0} D^n f \right) = v_0((I - D)^{-1} f);$$

thus if R verifies the equation

$$(I - D)R = f, \tag{1}$$

then we get

$$\sum_{n \geq 0} a_n = v_0(R)$$

Of course, such an algebraic definition of summation needs some hypotheses, especially to assure uniqueness of the solution of Equation (1); this is presented in Chap. 5.

It is easy to see that the Cauchy summation is a special case of this algebraic formalism: we take E as the vector space of convergent complex sequences $u = (u_n)_{n \geq 0}$ and

$$D : (u_n) \mapsto (u_{n+1})$$

$$v_0 : (u_n) \mapsto u_0$$

In this case the generator of a sequence of complex numbers $(a_n)_{n \geq 0}$ is precisely this sequence $f = (a_n)_{n \geq 0}$ since $(D^n f)_k = a_{k+n}$. If we set $R = (r_n)_{n \geq 0}$, Eq. (1) becomes the difference equation

$$r_n - r_{n+1} = a_n$$

The solution of this equation is defined up to an arbitrary constant; thus to get a unique solution, we need to impose a condition on $(r_n)_{n \geq 0}$. Since

$$r_0 - r_n = a_0 + \ldots + a_{n-1},$$

we see that if we add the condition

$$\lim_{n \to +\infty} r_n = 0,$$

then we have a unique solution R of (1) and we get the usual convergence of the series $\sum_{n\geq0} a_n$ with

$$v_0(R) = r_0 = \lim_{n\to+\infty} (a_0 + \ldots + a_{n-1}) = \sum_{n=0}^{+\infty} a_n$$

With this algebraic setting, we have presented (Candelpergher 1995) the summation method employed by Ramanujan in Chapter VI of his second notebook (the notebooks of Ramanujan are edited by the Tata Institute of Fundamental Research). This is done in a way similar to the Cauchy summation but replacing the space of sequences by a certain space E of analytic functions and

$$Df(x) = f(x+1)$$

$$v_0(f) = f(1)$$

For the Ramanujan summation, the terms of the series are indexed with $n \geq 1$ as Ramanujan does, and since $v_0(D^n f) = f(n)$, we define this summation for the series of type $\sum_{n\geq1} f(n)$ where $f \in E$.

In this case Eq. (1) is simply the difference equation

$$R(x) - R(x+1) = f(x) \tag{2}$$

If we can select a solution R_f of this equation, then we define the Ramanujan summation by

$$\sum_{n\geq1}^{\mathcal{R}} f(n) = R_f(1) \tag{3}$$

Existence and uniqueness of a solution of (2) can be proved by use of a general Laplace transform (Candelpergher et al. 1997).

Now, in Chap. 1, we give a more simple presentation that avoids the Laplace transform, and we give more explicit formulas. We start, as Ramanujan in his second notebook, with the summation formula of Euler and MacLaurin

$$f(1) + \ldots + f(n) = C_f + \int_1^n f(x)dx + f(n) + \sum_{k\geq1} \frac{B_k}{k!} \partial^{k-1} f(n)$$

This formula can be viewed as an asymptotic expansion, when x goes to infinity, of the function that Ramanujan writes

$$\varphi_f : x \mapsto f(1) + \ldots + f(x)$$

which is a hypothetical interpolation function of the partial sums of the series $\sum_{n\geq1}f(n)$, given with the condition $\varphi_f(0) = 0$. This expansion contains a constant C_f that Ramanujan calls "the constant of the series" and treats like a sort of sum for the series.

Since $\varphi_f(x + 1) - \varphi_f(x) = f(x + 1)$, we see that if we set

$$R_f(x) = C_f + f(x) - \varphi_f(x),$$

then the function R_f is a solution of Equation (2) and

$$\sum_{n\geq1}^{\mathcal{R}}f(n) = R_f(1) = C_f$$

We use the term *Ramanujan constant* for $R_f(1)$ since the expression "Ramanujan sum" is widely used for another notion in number theory.

We give a precise definition of R_f in Chap. 1 by the use of an integral formula which is related to the Abel-Plana summation formula.

For a function f sufficiently decreasing, the Euler-MacLaurin summation formula gives

$$C_f = \lim_{n\to+\infty} \left(f(1) + \ldots + f(n) - \int_1^n f(x)dx\right) \tag{4}$$

Thus if we are in a case of convergence of the series and integral, then we have

$$\sum_{n\geq1}^{\mathcal{R}}f(n) = \sum_{n=1}^{+\infty}f(n) - \int_1^{+\infty} f(x)dx \tag{5}$$

There is also the apparition of an integral when we compare the sums $\sum_{n\geq1}^{\mathcal{R}}f(n)$ and $\sum_{n\geq1}^{\mathcal{R}}f(n+1)$, since by (4) we have, in contrast to the usual translation property, the unusual *shift property*

$$\sum_{n\geq1}^{\mathcal{R}}f(n + 1) = \sum_{n\geq1}^{\mathcal{R}}f(n) - f(1) + \int_1^2 f(x)dx \tag{6}$$

When we apply this property to the function R_f and use Eq. (2), we get

$$\int_1^2 R_f(x)dx = 0 \tag{7}$$

This gives a way to define a general Ramanujan summation, independently of the Euler-MacLaurin formula, by Eqs. (2) and (3) with R_f of exponential growth $< \pi$ and verifying the integral condition (7) in order to determine a unique solution R_f.

The shift property (6) remains valid in this general setting and is very useful; for example, the "Casimir series" $\sum_{n\geq1}(n-1)^k$ (with k a positive integer) are Ramanujan summables, and we have

$$\sum_{n\geq1}^{\mathcal{R}} n^k = \sum_{n\geq1}^{\mathcal{R}}(n-1)^k + \int_1^2 (x-1)^k dx = \sum_{n\geq1}^{\mathcal{R}}(n-1)^k + \frac{1}{k+1} \qquad (8)$$

We see that the series $1+2^k+3^k+4^k+\ldots$ and $0+1+2^k+3^k+4^k+\ldots$ don't have the same sum! This is a consequence of the fact that the Ramanujan summation of $\sum_{n\geq1}f(n)$ is intimately related to the function f (in the first case, we have $f(x)=x^k$ and, in the second, $f(x)=(x-1)^k$).

In Chap. 2 we study elementary properties of the Ramanujan summation in comparison to the classical properties of the usual Cauchy summation. We prove a surprising relation between the sums

$$\sum_{n\geq1}^{\mathcal{R}} \varphi_f(n) \text{ and } \sum_{n\geq1}^{\mathcal{R}} \int_1^n f(x)dx$$

We also give, in the special case of an entire function f of exponential type $< \pi$, a simple formula for the Ramanujan summation of a series $\sum_{n\geq1}f(n)$ in terms of a convergent series involving the Bernoulli numbers

$$\sum_{n\geq1}^{\mathcal{R}} f(n) = \int_0^1 f(x)dx - \frac{1}{2}f(0) - \sum_{k=1}^{+\infty} \partial^k f(0)\frac{B_{k+1}}{(k+1)!}$$

In Chap. 3 we give important theorems that concern properties of sums $\sum_{n\geq1}^{\mathcal{R}} f(z,n)$ where z is a parameter, with respect to derivation, integration, or summation of f in z.

The simplest introduction of an external parameter in a series $\sum_{n\geq1}f(n)$ is to consider the series $\sum_{n\geq1}f(n)e^{-nz}$. We prove that if f is a function of moderate growth, then we have

$$\sum_{n\geq1}^{\mathcal{R}} f(n) = \lim_{z\to0}\sum_{n\geq1}^{\mathcal{R}} f(n)e^{-nz}$$

But if $z > 0$, then the series $\sum_{n\geq1}f(n)e^{-nz}$ is convergent, and by (5) we have

$$\sum_{n\geq1}^{\mathcal{R}} f(n) = \lim_{z\to0+}\left(\sum_{n=1}^{+\infty}f(n)e^{-nz} - \int_1^{+\infty} f(x)e^{-zx}dx\right) \qquad (9)$$

which seems to be a standard method of regularization used in physics.

A very important property of the Ramanujan summation is that the analyticity of the terms of a series implies the analyticity of the sum. A simple example is given by the zeta series $\sum_{n\geq1}\frac{1}{n^z}$. This series is convergent only for $Re(z)>1$, but the Ramanujan summation of this series is defined for all $z\in\mathbb{C}$, and by (5) we have for $Re(z)>1$

$$\sum_{n\geq1}^{\mathcal{R}}\frac{1}{n^z}=\sum_{n\geq1}^{+\infty}\frac{1}{n^z}-\frac{1}{z-1}=\zeta(z)-\frac{1}{z-1}$$

By this property of analyticity, this formula remains valid for all $z\neq1$, and we see that the Ramanujan summation of the zeta series cancels the pole of the zeta function at $z=1$. More precisely at $z=1$, we have by (4)

$$\sum_{n\geq1}^{\mathcal{R}}\frac{1}{n}=\lim_{n\to+\infty}\left(1+\ldots+\frac{1}{n}-Log(n)\right)=\gamma\quad\text{(Euler constant)}$$

We also note that for the "Casimir series" $\sum_{n\geq1}(n-1)^k$ (with k a positive integer), we have by (8)

$$\sum_{n\geq1}^{\mathcal{R}}(n-1)^k=\sum_{n\geq1}^{\mathcal{R}}n^k-\frac{1}{k+1}=\zeta(-k)\qquad(10)$$

In the second section of Chap. 3, we study the possibility to integrate the sum $\sum_{n\geq1}^{\mathcal{R}}f(z,n)$ with respect to z. For example, we have for $0<Re(s)<1$

$$\int_0^{+\infty}u^{s-1}\sum_{n\geq1}^{\mathcal{R}}\frac{1}{n+u}\,du=\sum_{n\geq1}^{\mathcal{R}}\int_0^{+\infty}u^{s-1}\frac{1}{n+u}\,du=\frac{\pi}{\sin\pi s}\sum_{n\geq1}^{\mathcal{R}}\frac{1}{n^{1-s}}$$

which gives simply the functional equation for the ζ function. Finally we give a sort of Fubini theorem for double sums

$$\sum_{m\geq1}^{\mathcal{R}}\sum_{n\geq1}^{\mathcal{R}}f(m,n)$$

We apply this result to the Eisenstein function G_2 (Freitag and Busam 2009)

$$G_2(z)=\sum_{n=-\infty}^{+\infty}\left(\sum_{\substack{m=-\infty\\m\neq0\,\text{if}\,n=0}}^{+\infty}\frac{1}{(m+nz)^2}\right)$$

Since we don't have absolute summability, we cannot interchange the sums, but thanks to the Ramanujan summation, we can prove simply that this function G_2 satisfies the nontrivial relation

$$G_2(-\frac{1}{z}) = z^2 G_2(z) - 2i\pi z$$

In Chap. 4 we give some relations of the Ramanujan summation to other summation formulas. First we recall the classical result (Hardy 1949) on the Borel summability of the asymptotic series $\sum_{k\geq 1} \frac{B_k}{k!} \partial^{k-1} f(n)$ that appears in the Euler-MacLaurin formula, and we give the formula linking $\sum_{n\geq 1}^{\mathcal{R}} f(n)$ with the Borel sum of the series $\sum_{k\geq 1} \frac{B_k}{k!} \partial^{k-1} f(1)$.

In the second section of this chapter, we use the Newton interpolation series (Nörlund 1926) to give a transformation formula involving a convergent series

$$\sum_{n\geq 1}^{\mathcal{R}} f(n) = \sum_{k=0}^{+\infty} \frac{\beta_{k+1}}{(k+1)!} (\Delta^k f)(1)$$

where the β_k are the Bernoulli numbers of the second kind and Δ is the usual difference operator. This formula is related to the classical Laplace summation formula (Boole 1960).

In the third section, we see that we can define the Euler summation of alternate series by using the Euler-Boole summation formula that is an analogue of the Euler-Maclaurin formula. We see that this summation is given by

$$\sum_{n\geq 1}^{\mathcal{E}} (-1)^{n-1} f(n) = \sum_{k=0}^{+\infty} \frac{(-1)^k}{2^{k+1}} (\Delta^k f)(1),$$

and we prove that this sum is related to the sums $\sum_{n\geq 1}^{\mathcal{R}} f(2n-1)$ and $\sum_{n\geq 1}^{\mathcal{R}} f(2n)$ by

$$\sum_{n\geq 1}^{\mathcal{R}} f(2n-1) - \sum_{n\geq 1}^{\mathcal{R}} f(2n) = \sum_{n\geq 1}^{\mathcal{E}} (-1)^{n-1} f(n) - \frac{1}{2} \int_1^2 f(t)dt$$

This can also be generalized to series of type $\sum_{n\geq 1} \omega^{n-1} f(n)$ where ω is a root of unity.

The unusual shift property (6) shows that the Ramanujan summation does not verify the translation property which is required (with linearity) for a summation method by Hardy in his very fine book *Divergent Series*. But we have seen that abandon of the translation property is necessary if we want to sum series like $\sum_{n\geq 1} P(n)$ where P is a polynomial. Thus, in Chap. 5, we give a general algebraic theory of summation of series that unifies the classical summation methods and the Ramanujan summation. In this general framework, the usual translation property appears as a special case of a more general shift property.

In the appendix, we give the classical Euler-MacLaurin and Euler-Boole formulas and a proof of Carlson's theorem.

Obviously we cannot claim that our version of the Ramanujan summation is exactly the summation procedure that Ramanujan had in mind, so the reader can find an exact copy of the Chapter VI of the second Notebook (in which Ramanujan introduces the "constant of a series") in the new edition of Ramanujan Notebooks published by the Tata Institute of Fundamental Research.

Acknowledgments

My warmest thanks go to M.A. Coppo, E. Delabaere, H. Gopalkrishna Gadiyar, and R. Padma for their interest and contributions to the study of the Ramanujan summation. I also wish to express my gratitude to F. Rouviere for his helpful comments and his constant encouragements.

Contents

Notation

- For $x = a + ib$ where a and b are real, we use the notation

$$a = Re(x) \text{ and } b = Im(x)$$

- The derivative of a function f is denoted by f' or ∂f.
- The difference operator is defined by $\Delta(f) = f(x+1) - f(x)$.
- For $k \in \mathbb{N}$ and $j = 0, \ldots, k$, we use the binomial coefficient

$$C_k^j = \frac{k!}{j!(k-j)!}$$

- The harmonic numbers are $H_n^{(s)} = 1 + \frac{1}{2^s} + \ldots + \frac{1}{n^s}$.
- The Bernoulli polynomials $B_k(x)$ are defined by

$$\sum_{n \geq 0} \frac{B_n(x)}{n!} t^n = \frac{t e^{xt}}{e^t - 1}$$

and the Bernoulli numbers $B_n = B_n(0)$
We have $B_0 = 1$, $B_1 = -1/2$, $B_{2n+1} = 0$ if $n \geq 1$.

- We define the Bernoulli numbers of the second kind β_n by

$$\frac{t}{\log(1+t)} = \sum_{n \geq 0} \frac{\beta_n}{n!} t^n = 1 + \sum_{n \geq 0} \frac{\beta_{n+1}}{(n+1)!} t^{n+1}$$

- The Euler polynomials $E_n(x)$ are defined by

$$\sum_{n \geq 0} \frac{E_n(x)}{n!} z^n = \frac{2 e^{xz}}{e^z + 1}$$

and we set $E_n = E_n(0)$.

- For $x \in \mathbb{C}$ we use the notation $\int_1^x f(u)du$ for the integral

$$\int_1^x f(u)du = \int_0^1 f(1 + t(x-1))(x-1)dt$$

  that is the integral on the segment relying 1 to x.
- Log is the logarithm function defined on $\mathbb{C}\backslash]-\infty, 0]$ by

$$Log(x) = \int_1^x \frac{1}{u}du$$

  that is for $x \in \mathbb{C}\backslash]-\infty, 0]$ by

$$Log(x) = Log(|x|) + i\theta \ \text{ if } x = |x|e^{i\theta}, \theta \in]-\pi, \pi[$$

- The Stieltjes constants γ_k are defined by

$$\gamma_k = \lim_{n \to +\infty} \Big(\sum_{j=1}^n \frac{Log^k(j)}{j} - \frac{Log^{k+1}(n)}{k+1} \Big)$$

  The Euler constant γ is γ_0 .
- The Catalan constant is $G = \sum_{n=1}^{+\infty} \frac{(-1)^{n-1}}{(2n-1)^2}$.
- The digamma function is

$$\psi(z) = \frac{\Gamma'(z)}{\Gamma(z)}$$

- We define $\mathcal{O}^\alpha$ the space of functions g analytic in a half plane

$$\{x \in \mathbb{C} | Re(x) > a\} \text{ with some } a < 1$$

  and of exponential type $< \alpha$: there exists $\beta < \alpha$ such that

$$|g(x)| \leq Ce^{\beta|x|} \text{ for } Re(x) > a$$

- We say that f is of *moderate growth* if f is analytic in a half plane

$$\{x \in \mathbb{C} | Re(x) > a\} \text{ with some } a < 1$$

  and of exponential type $< \varepsilon$ for all $\varepsilon > 0$.
- For $f \in \mathcal{O}^\pi$ the function R_f is the unique solution in $\mathcal{O}^\pi$ of

$$R_f(x) - R_f(x+1) = f(x) \text{ with } \int_1^2 R_f(x)dx = 0$$

- For $f \in \mathcal{O}^\pi$ the function φ_f is the unique function in $\mathcal{O}^\pi$ such that

$$\varphi_f(n) = f(1) + \ldots + f(n) \text{ for every integer } n \geq 1$$

- For $f \in \mathcal{O}^\pi$ the *Ramanujan summation* of $\sum_{n\geq 1} f(n)$ is defined by

$$\sum_{n\geq 1}^{\mathcal{R}} f(n) = R_f(1)$$

If the series is convergent, then $\sum_{n=1}^{+\infty} f(n)$ denotes its usual sum.

Chapter 1
Ramanujan Summation

In the first two sections of this chapter we recall the Euler-MacLaurin formula and use it to define what Ramanujan, in Chapter VI of his second Notebook, calls the "constant" of a series. But, as Hardy has observed, Ramanujan leaves some ambiguity in the definition of this "constant". Thus in the third section we interpret this constant as the value of a precise solution of a difference equation. Then we can give in Sect. 1.4 a rigorous definition of the Ramanujan summation and its relation to the usual summation for convergent series.

1.1   The Euler-MacLaurin Summation Formula

Let's consider a function $f \in C^{\infty}(]0, +\infty[)$. For every positive integer n we can write

$$\sum_{k=1}^{n} f(k) = \sum_{k=1}^{n} (k - (k - 1)) f(k)$$

$$= nf(n) - \sum_{k=1}^{n-1} k(f(k + 1) - f(k))$$

$$= nf(n) - \sum_{k=1}^{n-1} \int_{k}^{k+1} [x] f'(x) dx$$

where $[x]$ is the integral part of x. Let $\{x\} = x - [x]$, then

$$\sum_{k=1}^{n} f(k) = nf(n) - \int_{1}^{n} x f'(x) dx + \int_{1}^{n} \{x\} f'(x) dx$$

© Springer International Publishing AG 2017
B. Candelpergher, *Ramanujan Summation of Divergent Series*,
Lecture Notes in Mathematics 2185, DOI 10.1007/978-3-319-63630-6_1

and integration by parts gives

$$\sum_{k=1}^{n} f(k) = f(1) + \int_1^n f(x)dx + \int_1^n \{x\}f'(x)dx$$

In order to generalize this formula we define the function

$$b_1(x) = \{x\} - \frac{1}{2}$$

which is a 1-periodic function with $\int_0^1 b_1(x)dx = 0$. We have

$$\sum_{k=1}^{n} f(k) = f(1) + \int_1^n f(x)dx + \int_1^n b_1(x)f'(x)dx + \frac{1}{2}\int_1^n f'(x)dx$$

thus we get the *Euler-MacLaurin formula to order* 1:

$$\sum_{k=1}^{n} f(k) = \int_1^n f(x)dx + \frac{1}{2}(f(n) + f(1)) + \int_1^n b_1(x)f'(x)dx \qquad (1.1)$$

To make another integration by parts in the last integral we introduce the function

$$c_2(x) = \int_0^x b_1(t)dt$$

which is a 1-periodic with $c_2(n) = 0$ for every positive integer n and

$$c_2(x) = \frac{x^2}{2} - \frac{x}{2} \text{ if } x \in]0, 1[$$

then by integration by parts

$$\int_1^n b_1(x)f'(x)dx = -\int_1^n c_2(x)f''(x)dx$$

Thus we get

$$\sum_{k=1}^{n} f(k) = \int_1^n f(x)dx + \frac{1}{2}(f(n) + f(1)) - \int_1^n c_2(x)f''(x)dx$$

We now replace the function c_2 by

$$\frac{b_2(x)}{2} = c_2(x) + \frac{B_2}{2}$$

with the choice of B_2 in order that

$$\int_0^1 \frac{b_2(x)}{2} dx = 0$$

which gives $B_2 = 1/6$ and $\frac{b_2(x)}{2} = \frac{x^2}{2} - \frac{x}{2} + \frac{1}{12}$ if $x \in [0, 1[$.

Then we get the *Euler-MacLaurin formula to order* 2:

$$\sum_{k=1}^n f(k) = \int_1^n f(x)dx + \frac{1}{2}(f(n) + f(1)) + \frac{B_2}{2}(f'(n) - f'(1)) - \int_1^n \frac{b_2(x)}{2} f''(x)dx$$

If we continue these integrations by parts we get a general *Euler-MacLaurin formula to order m*:

$$f(1) + \ldots + f(n) = \int_1^n f(x)dx + \frac{f(1) + f(n)}{2}$$

$$+ \sum_{k=2}^m \frac{(-1)^k B_k}{k!} [\partial^{k-1} f(n) - \partial^{k-1} f(1)]$$

$$+ (-1)^{m+1} \int_1^n \frac{b_m(x)}{m!} \partial^m f(x)dx$$

where $b_m(x)$ is the 1-periodic function $b_m(x) = B_m(x - [x])$, with the *Bernoulli polynomials* $B_k(x)$ defined by

$$\sum_{n \geq 0} \frac{B_n(x)}{n!} t^n = \frac{te^{xt}}{e^t - 1}$$

and the *Bernoulli numbers* $B_n = B_n(0)$ (we verify that $B_{2k+1} = 0$ for $k \geq 1$). For a simple proof of this general formula see Appendix.

1.2 Ramanujan's Constant of a Series

Let f be a C^∞ function defined for real $x > 0$. In the beginning of Chapter VI of his Notebook 2, Ramanujan introduces the hypothetical sum

$$f(1) + f(2) + f(3) + f(4) + \ldots \ldots + f(x) = \varphi(x),$$

which is intended to be the solution of

$$\varphi(x) - \varphi(x - 1) = f(x) \text{ with } \varphi(0) = 0$$

Let's take the numbers $\mathcal{B}_r$ defined when $r = 2, 4, 6, \ldots$ by (second notebook chapter V, entry 9)

$$\frac{x}{e^x - 1} = 1 - \frac{x}{2} + \sum_{k \geq 1} \frac{(-1)^{k-1} \mathcal{B}_{2k}}{(2k)!} x^{2k}$$

then Ramanujan writes the Euler-McLaurin series

$$\varphi(x) = C + \int f(x)dx + \frac{1}{2}f(x) + \frac{\mathcal{B}_2}{\lfloor 2} f'(x) - \frac{\mathcal{B}_4}{\lfloor 4} f'''(x) + \frac{\mathcal{B}_6}{\lfloor 6} f^V(x) + \frac{\mathcal{B}_8}{\lfloor 8} f^{VII}(x) + \ldots$$

and he says about the constant C: *the algebraic constant of a series is the constant obtained by completing the remaining part in the above theorem. We can substitute this constant which is like the centre of gravity of a body instead of its divergent infinite series.*

In Ramanujan's notation the numbers $\mathcal{B}_{2n}$ are related with the usual Bernoulli numbers by

$$\mathcal{B}_{2n} = (-1)^{n-1} B_{2n}$$

Thus we can write the above Euler-McLaurin series in the form

$$\varphi(x) = C + \int f(x)dx + \frac{1}{2}f(x) + \sum_{k \geq 2} \frac{B_{2k}}{(2k)!} \partial^{2k-1} f(x)$$

The main difficulty with this formula is that this last series is not always convergent. Therefore we replace this series by a finite sum and give a precise meaning to the integral. We are thus led to write the Euler-MacLaurin summation formula in the form

$$f(1) + \ldots + f(n) = C_m(f) + \int_1^n f(x)dx + \frac{f(n)}{2} + \sum_{k=1}^m \frac{B_{2k}}{(2k)!} \partial^{2k-1} f(n)$$

$$- \int_n^{+\infty} \frac{b_{2m+1}(x)}{(2m+1)!} \partial^{2m+1} f(x)dx$$

where

$$C_m(f) = \frac{f(1)}{2} - \sum_{k=1}^m \frac{B_{2k}}{(2k)!} \partial^{2k-1} f(1) + \int_1^{+\infty} \frac{b_{2m+1}(x)}{(2m+1)!} \partial^{2m+1} f(x)dx$$

in this formula we assume that the function f is an infinitely differentiable function and that the integral

$$\int_1^{+\infty} b_{2m+1}(x)\partial^{2m+1}f(x)dx$$

is convergent for all $m \geq M > 0$. Then by integration by parts we verify that the constant $C_m(f)$ does not depend on m if $m \geq M$ thus we set $C_m(f) = C(f)$.

We use the notation

$$C(f) = \sum_{n\geq 1}^{\mathcal{R}} f(n)$$

and call it the *Ramanujan's constant of the series*.

Example If f is a constant function then $\partial f = 0$ thus $\sum_{n\geq 1}^{\mathcal{R}} f(n) = \frac{f(1)}{2}$. Thus

$$\sum_{n\geq 1}^{\mathcal{R}} 1 = \frac{1}{2}$$

If $f(x) = x$ then $\partial^2 f = 0$ thus $\sum_{n\geq 1}^{\mathcal{R}} n = \frac{1}{2} - \frac{B_2}{2} = \frac{5}{12}$.

The Case of Convergence

Assume that the integral $\int_1^{+\infty} b_1(x)\partial f(x)dx$ is convergent, then the Euler-MacLaurin formula to order 1 is simply

$$f(1) + \ldots + f(n) = C(f) + \int_1^n f(x)dx + \frac{f(n)}{2} - \int_n^{+\infty} b_1(x)f'(x)dx$$

with

$$C(f) = \frac{f(1)}{2} + \int_1^{+\infty} b_1(x)f'(x)dx$$

Since $\int_n^{+\infty} b_1(x)f'(x)dx \to 0$ when $n \to +\infty$ we get the following expression of the Ramanujan constant as the limit

$$\sum_{n\geq 1}^{\mathcal{R}} f(n) = \lim_{n\to+\infty} \left(f(1) + \ldots + f(n) - \int_1^n f(x)dx - \frac{f(n)}{2} \right) \qquad (1.2)$$

If in addition we assume that $\lim_{n\to+\infty} f(n) = 0$ then we have

$$\sum_{n\geq 1}^{\mathcal{R}} f(n) = \lim_{n\to+\infty} \left(f(1) + \ldots + f(n-1) - \sum_{k=1}^{n-1} \int_k^{k+1} f(x)dx \right)$$

thus we get an expression of the Ramanujan constant as a sum of a convergent series

$$\sum_{n\geq 1}^{\mathcal{R}} f(n) = \sum_{n=1}^{+\infty} \left(f(n) - \int_{n}^{n+1} f(x)dx \right) \tag{1.3}$$

This proves the property that *if the series* $\sum_{n\geq 1} f(n)$ *and the integral* $\int_{1}^{+\infty} f(x)dx$ *are convergent* then we have the relation

$$\sum_{n\geq 1}^{\mathcal{R}} f(n) = \sum_{n\geq 1}^{+\infty} f(n) - \int_{1}^{+\infty} f(x)dx \tag{1.4}$$

Examples

(1) If $f(x) = \frac{1}{x^z}$ with $Re(z) > 1$ then by (1.4)

$$\sum_{n\geq 1}^{\mathcal{R}} \frac{1}{n^z} = \sum_{n\geq 1}^{+\infty} \frac{1}{n^z} - \int_{1}^{+\infty} \frac{1}{x^z}dx = \sum_{n\geq 1}^{+\infty} \frac{1}{n^z} - \frac{1}{z-1}$$

thus for $Re(z) > 1$ we have the relation with the classical Riemann zeta function

$$\sum_{n\geq 1}^{\mathcal{R}} \frac{1}{n^z} = \zeta(z) - \frac{1}{z-1} \tag{1.5}$$

(2) If $f(x) = \frac{1}{x}$ then by (1.2)

$$\sum_{n\geq 1}^{\mathcal{R}} \frac{1}{n} = \lim_{n\to +\infty} \left(\sum_{k=1}^{n} \frac{1}{k} - Log(n) - \frac{1}{2n} \right) = \lim_{n\to +\infty} \left(\sum_{k=1}^{n} \frac{1}{k} - Log(n) \right)$$

thus

$$\sum_{n\geq 1}^{\mathcal{R}} \frac{1}{n} = \gamma \text{ where } \gamma \text{ is the Euler constant} \tag{1.6}$$

(3) If $f(x) = Log(x)$ then by (1.2)

$$\sum_{n\geq 1}^{\mathcal{R}} Log(n) = \lim_{n\to +\infty} \left(\sum_{k=1}^{n} Log(k) - (nLog(n) - n + 1 + \frac{1}{2}Log(n)) \right)$$

Using the Stirling formula we have

$$\lim_{n\to+\infty}\sum_{k=1}^{n}Log(k) - (nLog(n) - n + \frac{1}{2}Log(n)) = Log(\sqrt{2\pi})$$

this gives

$$\sum_{n\geq1}^{\mathcal{R}}Log(n) = Log(\sqrt{2\pi}) - 1 \tag{1.7}$$

(4) If $f(x) = xLog(x)$ then we have $\partial f(x) = Log(x) + 1$ and $\partial^2 f(x) = \frac{1}{x}$. Thus by the preceding Euler-MacLaurin formula with $m = 1$ we have

$$\sum_{n\geq1}^{\mathcal{R}}nLog(n) = \lim_{n\to+\infty}\left(\sum_{k=1}^{n}kLog(k) - \int_{1}^{n}xLog(x)dx - \frac{nLog(n)}{2} - \frac{Log(n)+1}{12}\right)$$

this gives

$$\sum_{n\geq1}^{\mathcal{R}}nLog(n) = \lim_{n\to+\infty}\left(\sum_{k=1}^{n}kLog(k) - Log(n)(\frac{n^2}{2} + \frac{n}{2} + \frac{1}{12}) + \frac{n^2}{4}\right) - \frac{1}{3}$$

thus

$$\sum_{n\geq1}^{\mathcal{R}}nLog(n) = Log(A) - \frac{1}{3} \tag{1.8}$$

where A is the Glaisher-Kinkelin constant (Srivastava and Junesang Choi 2012, p.39).

This constant is related to the zeta function by the following relation that we prove later on

$$\sum_{n\geq1}^{\mathcal{R}}nLog(n) = -\zeta'(-1) - \frac{1}{4}$$

Remark Note that with the Euler-MacLaurin formula we have for all $a > 0$:

$$f(1) + \ldots + f(n) =$$

$$\int_{1}^{a}f(x)dx + \frac{f(1)}{2} - \sum_{k=1}^{m}\frac{B_{2k}}{(2k)!}\partial^{2k-1}f(1) + \int_{1}^{+\infty}\frac{b_{2m+1}(x)}{(2m+1)!}\partial^{2m+1}f(x)dx$$

$$+ \int_{a}^{n}f(x)dx + \frac{f(n)}{2} + \sum_{k=1}^{m}\frac{B_{2k}}{(2k)!}\partial^{2k-1}f(n) - \int_{n}^{+\infty}\frac{b_{2m+1}(x)}{(2m+1)!}\partial^{2m+1}f(x)dx$$

This is a summation formula where the constant $C(f) = \sum_{n\geq1}^{\mathcal{R}} f(n)$ is replaced by

$$C_a(f) = \int_1^a f(x)dx + \sum_{n\geq1}^{\mathcal{R}} f(n)$$

It seems that Ramanujan leaves the possibility that the choice of a depends on the series considered.

In the special case where the series $\sum f(n)$ and the integral $\int_1^{+\infty} f(x)dx$ are convergent then if we take $a = +\infty$ we get

$$C_\infty(f) = \int_1^{+\infty} f(x)dx + \sum_{n\geq1}^{\mathcal{R}} f(n)$$

and with the relation (1.4) we get

$$C_\infty(f) = \sum_{n=1}^{+\infty} f(n)$$

This explains an assertion of Ramanujan in Notebook 2 Chapter 6 p.62:

"*If $f(1) + f(2) + \ldots + f(x)$ be a convergent series then its constant is the sum of the series.*"

But if for example $f(x) = x$, then it is not possible to take $a = +\infty$ since $C_\infty(f)$ is not defined but if we take for example $a = 0$ then

$$C_0(f) = -\int_0^1 x\,dx + \sum_{n\geq1}^{\mathcal{R}} n = -\frac{B_2}{2}$$

To get simple properties of the Ramanujan summation we fix the parameter a in the integral, and we make the choice $a = 1$ in order to have

$$\sum_{n\geq1}^{\mathcal{R}} \frac{1}{n} = \gamma$$

Conclusion

With the use of Euler-Maclaurin formula we have the definition of the constant of a series by

$$\sum_{n\geq1}^{\mathcal{R}} f(n) = \lim_{n\to+\infty}(f(1) + \ldots + f(n)) - [\int_1^n f(x)dx + \frac{f(n)}{2} + \sum_{k=1}^m \frac{B_{2k}}{(2k)!}\partial^{2k-1}f(n)]$$

$$(1.9)$$

this needs convergence of the integral

$$\int_1^{+\infty} b_{2m+1}(x)\partial^{2m+1}f(x)dx$$

This last hypothesis is not always satisfied, for example if we look at a series like $\sum_{n\geq 1} e^n$. Thus we need to avoid the systematic use of Euler-McLaurin formula and define in a more algebraic way the Ramanujan summation.

1.3 A Difference Equation

1.3.1 The Functions φ_f and R_f

In his Notebook Ramanujan uses the function φ formally defined by

$$\varphi(x) = f(1) + \ldots + f(x)$$

It seems he has in mind a *sort of unique interpolation function* φ_f of the partial sums $f(1) + f(2) + \ldots + f(n)$ of the series $\sum_{n\geq 1} f(n)$ associated to f. This interpolation function must verify

$$\varphi_f(x) - \varphi_f(x-1) = f(x) \tag{1.10}$$

and Ramanujan sets the additional condition $\varphi_f(0) = 0$. With this condition the relation (1.10) gives for every n integer ≥ 1

$$\varphi_f(n) = f(1) + f(2) + \ldots + f(n)$$

Note that if the series $\sum_{n\geq 1} f(n)$ is convergent we have

$$\lim_{n\to+\infty} \varphi_f(n) = \sum_{n=1}^{+\infty} f(n)$$

Now in general the Euler-Mclaurin summation formula gives an expansion of the function φ_f which we can write

$$\varphi_f(n) = C(f) + f(n) - R_f(n)$$

where the function R_f is defined by

$$R_f(n) = \frac{f(n)}{2} - \sum_{k=1}^{m} \frac{B_{2k}}{(2k)!}\partial^{2k-1}f(n) + \int_n^{+\infty} \frac{b_{2m+1}(t)}{(2m+1)!}\partial^{2m+1}f(t)dt - \int_1^n f(t)dt \tag{1.11}$$

and

$$C(f) = \frac{f(1)}{2} - \sum_{k=1}^{m} \frac{B_{2k}}{(2k)!} \partial^{2k-1} f(1) + \int_{1}^{+\infty} \frac{b_{2m+1}(x)}{(2m+1)!} \partial^{2m+1} f(x) dx$$

We observe that $C(f) = R_f(1)$. Thus we get

$$\sum_{n\geq 1}^{\mathcal{R}} f(n) = R_f(1) \tag{1.12}$$

1.3.2   The Fundamental Theorem

To avoid the systematic use of the Euler-MacLaurin summation formula we now find another way to define the function R_f. The difference equation

$$\varphi_f(n+1) - \varphi_f(n) = f(n+1)$$

and the relation $\varphi_f(n) = C(f) + f(n) - R_f(n)$ gives for R_f the difference equation

$$R_f(n) - R_f(n+1) = f(n)$$

By (1.12) it seems natural to define the Ramanujan summation of the series $\sum_{n\geq 1} f(n)$ by

$$\sum_{n\geq 1}^{\mathcal{R}} f(n) = R(1)$$

where the function R satisfies the difference equation

$$R(x) - R(x+1) = f(x)$$

Clearly this equation is not sufficient to determine the function R, so we need additional conditions on R. Let us try to find these conditions.

First we see, by the definition (1.11) of R_f, that if f and its derivatives are sufficiently decreasing at $+\infty$ then we have

$$\lim_{n\to +\infty} R_f(n) = -\int_{1}^{+\infty} f(x) dx$$

But we can't impose this sort of condition on the function R because it involves the integral $\int_1^{+\infty} f(x)dx$ which, in the general case, is divergent. Thus we now translate it into another form.

Suppose we have a smooth function R solution of the difference equation

$$R(x) - R(x+1) = f(x) \text{ for all } x > 0$$

Now if we integrate the two sides of this equation between k and $k+1$ for all integer $k \geq 1$, and formally take the infinite sum over k, we obtain

$$\int_1^{+\infty} f(x)dx = \int_1^2 R(x)dx - \lim_{x \to +\infty} R(x)$$

We see that for the function R we have the equivalence

$$\lim_{x \to +\infty} R(x) = -\int_1^{+\infty} f(x)dx \text{ if and only if } \int_1^2 R(x)dx = 0$$

Thus we can try to define the function R_f by the difference equation $R_f(x) - R_f(x+1) = f(x)$ with the condition

$$\int_1^2 R_f(x)dx = 0$$

Unfortunately this does not define a unique function R_f because we can add to R_f any combination of periodic functions $x \mapsto e^{2i\pi kx}$. To avoid this problem we add the hypothesis that R_f is analytic in the half plane $\{x \in \mathbb{C} | Re(x) > 0\}$ and of exponential type $< 2\pi$:

Definition 1 A function g analytic for $Re(x) > a$ is of exponential type $< \alpha$ ($\alpha > 0$) if there exists $\beta < \alpha$ such that

$$|g(x)| \leq Ce^{\beta|x|} \text{ for } Re(x) > a$$

We define $\mathcal{O}^\alpha$ the space of functions g analytic in a half plane

$$\{x \in \mathbb{C} | Re(x) > a\} \text{ with some } a < 1$$

and of exponential type $< \alpha$ in this half plane.

We say that f is of *moderate growth* if f is analytic in this half plane and of exponential type $< \varepsilon$ for all $\varepsilon > 0$.

With this definition we have the following lemma:

Lemma 1 (Uniqueness Lemma) *Let* $R \in \mathcal{O}^{2\pi}$, *be a solution of*

$$R(x) - R(x+1) = 0 \text{ with } \int_1^2 R(x)dx = 0$$

then $R = 0$.

Proof By the condition $R(x) - R(x+1) = 0$, we see that R can be extended to an entire function. And we can write

$$R(x) = R_0(e^{2i\pi x})$$

where R_0 is the analytic function in $\mathbb{C}\setminus\{0\}$ given by

$$R_0(z) = R(\frac{1}{2i\pi}Log(z))$$

(where Log is defined by $Log(re^{i\theta}) = \ln(r) + i\theta$ with $0 \leq \theta < 2\pi$).

The Laurent expansion of R_0 gives

$$R(x) = \sum_{n\in\mathbb{Z}} c_n e^{2i\pi nx}$$

the coefficients c_n are

$$c_n = \frac{1}{2\pi r^n} \int_0^{2\pi} R_0(re^{it})e^{-int}dt = \frac{1}{2\pi r^n} \int_0^{2\pi} R(\frac{t}{2\pi} + \frac{1}{2i\pi}\ln(r))e^{-int}dt$$

where $r > 0$.

The condition that R is of exponential type $< 2\pi$ gives

$$|c_n| \leq \frac{1}{r^n}Ce^{\frac{\alpha}{2\pi}|\ln(r)|} \text{ with } \frac{\alpha}{2\pi} < 1.$$

If we take $r \to 0$ we get $c_n = 0$ for $n < 0$ and if we take $r \to +\infty$ then we get $c_n = 0$ for $n > 0$. Finally the condition $\int_1^2 R(x)dx = 0$ then gives $c_0 = 0$.   □

Theorem 1 *If $f \in \mathcal{O}^\alpha$ with $\alpha \leq 2\pi$ there exists a unique function $R_f \in \mathcal{O}^\alpha$ such that $R_f(x) - R_f(x+1) = f(x)$ with $\int_1^2 R_f(x)dx = 0$. This function is*

$$R_f(x) = -\int_1^x f(t)dt + \frac{f(x)}{2} + i\int_0^{+\infty} \frac{f(x+it) - f(x-it)}{e^{2\pi t} - 1}dt \qquad (1.13)$$

Proof

(a) Uniqueness is given by the preceding lemma.
(b) The function R_f defined by (1.13) is clearly in $\mathcal{O}^\alpha$.
(c) Let us prove that $R_f(x) - R_f(x+1) = f(x)$. By analyticity it is sufficient to prove this for real x.

Consider the integral

$$\int_\gamma f(z)\frac{1}{2i}\cot(\pi(z-x))dz$$

with γ the path

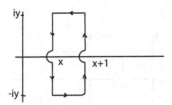

By the residue theorem we have

$$\int_{\gamma} f(z) \frac{1}{2i} \cot(\pi(z-x))dz = f(x)$$

To evaluate the different contributions of the integral we use the formulas:

$$\frac{1}{2i} \cot(\pi(z-x)) = -\frac{1}{2} - \frac{1}{e^{-2i\pi(z-x)}-1} \quad \text{when } Im(z) > 0$$

and

$$\frac{1}{2i} \cot(\pi(z-x)) = \frac{1}{2} + \frac{1}{e^{2i\pi(z-x)}-1} \quad \text{when } Im(z) < 0.$$

Let us examine the different contributions of the integral:

* the semicircular path at x and $x+1$ gives when $\varepsilon \to 0$

$$\frac{1}{2}f(x) - \frac{1}{2}f(x+1)$$

* the horizontal lines give

$$-(-\frac{1}{2})\int_{x}^{x+1} f(t+iy)dt + \frac{1}{2}\int_{x}^{x+1} f(t-iy)dt$$

and two additional terms which vanish when $y \to +\infty$ (by the hypothesis that f of exponential type $< 2\pi$).
* the vertical lines give

$$i\int_{\varepsilon}^{y} \frac{f(x+it)-f(x-it)}{e^{2\pi t}-1}dt - i\int_{\varepsilon}^{y} \frac{f(x+1+it)-f(x+1-it)}{e^{2\pi t}-1}dt$$

and

$$\frac{1}{2}\int_{\varepsilon}^{y} f(x+it)idt - \frac{1}{2}\int_{\varepsilon}^{y} f(x-it)idt - \frac{1}{2}\int_{\varepsilon}^{y} f(x+1+it)idt + \frac{1}{2}\int_{\varepsilon}^{y} f(x+1-it)idt$$

If we add this term with the contributions of the horizontal lines we obtain the sum of the integrals of f on the paths

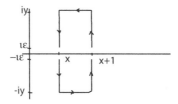

By the Cauchy theorem this sum is

$$\frac{1}{2}\int_x^{x+1} f(t+i\varepsilon)dt + \frac{1}{2}\int_x^{x+1} f(t-i\varepsilon)dt$$

which gives the contribution $\int_x^{x+1} f(t)dt$ when $\varepsilon \to 0$.
Finally when $\varepsilon \to 0$ and $y \to +\infty$ we get

$$f(x) = \frac{1}{2}f(x) - \frac{1}{2}f(x+1)$$
$$+i\int_0^{+\infty} \frac{f(x+it) - f(x-it)}{e^{2\pi t} - 1}dt$$
$$-i\int_0^{+\infty} \frac{f(x+1+it) - f(x+1-it)}{e^{2\pi t} - 1}$$
$$+\int_x^{x+1} f(t)dt$$

This is $f(x) = R_f(x) - R_f(x+1)$ with R_f given by (1.13).

(d) It remains to prove that $\int_1^2 R_f(x)dx = 0$. By Fubini's theorem

$$\int_1^2 \int_0^{+\infty} \frac{f(x+it) - f(x-it)}{e^{2\pi t} - 1}dt\,dx = \int_0^{+\infty} \int_1^2 \frac{f(x+it) - f(x-it)}{e^{2\pi t} - 1}dx\,dt$$

We have for $1 < x < 2$

$$\int_1^2 (f(x+it) - f(x-it))dx = F(2+it) - F(2-it)$$
$$- (F(1+it) - F(1-it))$$

where $F(x) = \int_1^x f(t)dt$. Thus

$$\int_1^2 \int_0^{+\infty} \frac{f(x+it) - f(x-it)}{e^{2\pi t} - 1} dt\, dx = \int_0^{+\infty} \frac{F(2+it) - F(2-it)}{e^{2\pi t} - 1} dt$$

$$- \int_0^{+\infty} \frac{F(1+it) - F(1-it)}{e^{2\pi t} - 1} dt$$

By the preceding result (applied with F in place of f) we have

$$F(x) = \frac{1}{2}F(x) - \frac{1}{2}F(x+1)$$

$$+ i \int_0^{+\infty} \frac{F(x+it) - F(x-it)}{e^{2\pi t} - 1} dt$$

$$- i \int_0^{+\infty} \frac{F(x+1+it) - F(x+1-it)}{e^{2\pi t} - 1}$$

$$+ \int_x^{x+1} F(t)dt$$

With $x = 1$ we get

$$i \int_1^2 \int_0^{+\infty} \frac{f(x+it) - f(x-it)}{e^{2\pi t} - 1} dt\, dx = -\frac{F(1) + F(2)}{2} + \int_1^2 F(t)dt$$

This gives

$$\int_1^2 R_f(x)dx = -\int_1^2 F(x)dx + \frac{1}{2}\int_1^2 f(x)dx - (\frac{F(1) + F(2)}{2}) + \int_1^2 F(t)dt = 0$$

$\square$

1.4 The Summation

1.4.1 Definition and Examples

Consider $f \in \mathcal{O}^{2\pi}$, then by Theorem 1 we can define the Ramanujan summation of the series $\sum_{n\geq 1} f(n)$ by

$$\sum_{n\geq 1}^{\mathcal{R}} f(n) = R_f(1)$$

where R_f is the unique solution in $\mathcal{O}^{2\pi}$ of

$$R_f(x) - R_f(x+1) = 0 \text{ with } \int_1^2 R_f(x)dx = 0.$$

We immediately note that for some $f \in \mathcal{O}^{2\pi}$ this definition can give surprising results. Let us consider for example the function $f : x \mapsto \sin(\pi x)$, since

$$\frac{\sin(\pi x)}{2} - \frac{\sin(\pi(x+1))}{2} = \sin(\pi x)$$

we get

$$R_f(x) = \frac{\sin(\pi x)}{2} - \int_1^2 \frac{\sin(\pi x)}{2} dx = \frac{\sin(\pi x)}{2} + \frac{1}{\pi}$$

thus

$$\sum_{n\geq 1}^{\mathcal{R}} \sin(\pi n) = \frac{1}{\pi}$$

But $\sin(\pi n) = g(n)$ with the function $g = 0$, and we have trivially $R_g = 0$ which gives

$$\sum_{n\geq 1}^{\mathcal{R}} 0 = 0$$

This example shows that for $f \in \mathcal{O}^{2\pi}$ the sum $\sum_{n\geq 1}^{\mathcal{R}} f(n)$ depends not only on the values $f(n)$ for integers $n \geq 1$ but also *on the interpolation function f* we have chosen.

To avoid this phenomenon we restrict the Ramanujan summation to functions f in $\mathcal{O}^{\pi}$, since with this condition we can apply Carlson's theorem (see Appendix) which says that such a function is uniquely determined by its values $f(n)$ for all integers $n \geq 1$. Note that in this case the function R_f given by Theorem 1 is also in $\mathcal{O}^{\pi}$.

Definition 2 If $f \in \mathcal{O}^{\pi}$, then there exist's a unique solution $R_f \in \mathcal{O}^{\pi}$ of

$$R_f(x) - R_f(x+1) = f(x) \text{ with } \int_1^2 R_f(x)dx = 0$$

We call this function R_f the *fractional remainder of f* and we set

$$\sum_{n\geq 1}^{\mathcal{R}} f(n) = R_f(1)$$

By (1.13) we have by the integral formula

$$\sum_{n\geq 1}^{\mathcal{R}} f(n) = \frac{f(1)}{2} + i \int_0^{+\infty} \frac{f(1+it) - f(1-it)}{e^{2\pi t} - 1} dt \qquad (1.14)$$

We call this procedure the *Ramanujan summation of* the series $\sum_{n\geq 1} f(n)$ and $\sum_{n\geq 1}^{\mathcal{R}} f(n)$ the *Ramanujan constant* of the series.

Some properties are immediate consequences of this definition:

Linearity

If a and b are complex numbers and f and g are in $\mathcal{O}^\pi$, then we verify immediately that

$$R_{af+bg} = aR_f + bR_g$$

thus the Ramanujan summation has the property of linearity

$$\sum_{n\geq 1}^{\mathcal{R}} af(n) + bg(n) = a \sum_{n\geq 1}^{\mathcal{R}} f(n) + b \sum_{n\geq 1}^{\mathcal{R}} g(n)$$

Reality

Consider $g \in \mathcal{O}^\pi$ such that $g(x) \in \mathbb{R}$ if $x \in \mathbb{R}$.

Then for all $t > 0$ we have by the reflection principle $g(1-it) = \overline{g(1+it)}$ thus $i(g(1+it) - g(1-it)) \in \mathbb{R}$ and by the integral formula (1.14) we have

$$\sum_{n\geq 1}^{\mathcal{R}} g(n) \in \mathbb{R}$$

Real and Imaginary Parts

Take $f \in \mathcal{O}^\pi$ defined in the half plane $\{x \in \mathbb{C}|Re(x) > a\}$ with $a < 1$. We define for $x \in]a, +\infty[$ the functions $f_r : x \mapsto Re(f(x))$ and $f_i : x \mapsto Im(f(x))$ and assume that *the functions f_r and f_i have analytic continuations on the half plane* $\{x \in \mathbb{C}|Re(x) > a\}$ *that are in $\mathcal{O}^\pi$*.

Then we can *define the sums* $\sum_{n\geq 1}^{\mathcal{R}} Re(f(n))$ and $\sum_{n\geq 1}^{\mathcal{R}} Im(f(n))$ by

$$\sum_{n\geq 1}^{\mathcal{R}} Re(f(n)) = \sum_{n\geq 1}^{\mathcal{R}} f_r(n)$$

$$\sum_{n\geq 1}^{\mathcal{R}} Im(f(n)) = \sum_{n\geq 1}^{\mathcal{R}} f_i(n)$$

By linearity we have

$$\sum_{n\geq 1}^{\mathcal{R}} f(n) = \sum_{n\geq 1}^{\mathcal{R}} (f_r(n)) + i f_i(n)) = \sum_{n\geq 1}^{\mathcal{R}} f_r(n) + i \sum_{n\geq 1}^{\mathcal{R}} f_i(n)$$

Since $f_r(x)$ and $f_i(x)$ are real for $x \in \mathbb{R}$ then by the reality property we get

$$Re(\sum_{n\geq 1}^{\mathcal{R}} f(n)) = \sum_{n\geq 1}^{\mathcal{R}} Re(f(n))$$

$$Im(\sum_{n\geq 1}^{\mathcal{R}} f(n)) = \sum_{n\geq 1}^{\mathcal{R}} Im(f(n))$$

For example if $f(z) = \frac{1}{z+i}$ then $f_r : z \mapsto \frac{z}{z^2+1}$ and $f_i : z \mapsto \frac{-1}{z^2+1}$ are analytic functions in $\mathcal{O}^\pi$. Thus

$$Re(\sum_{n\geq 1}^{\mathcal{R}} \frac{1}{n+i}) = \sum_{n\geq 1}^{\mathcal{R}} \frac{n}{n^2+1}$$

$$Im(\sum_{n\geq 1}^{\mathcal{R}} \frac{1}{n+i}) = \sum_{n\geq 1}^{\mathcal{R}} \frac{-1}{n^2+1}$$

Remark Note that generally we can't write $\sum_{n\geq 1}^{\mathcal{R}} \bar{f}(n) = \overline{\sum_{n\geq 1}^{\mathcal{R}} f(n)}$ since the function $\bar{f}$ is not analytic (if f is non constant).

Examples

(1) Take $f(x) = e^{-zx}$ with $z \in \mathbb{C}$ then for $Re(x) > 0$ we have

$$|f(x)| = e^{-Re(zx)} = e^{-Re(z\frac{x}{|x|})|x|}$$

Thus if we set $\frac{x}{|x|} = e^{i\theta}$ then $f \in \mathcal{O}^\pi$ for $z \in U_\pi$ where

$$U_\pi = \{z \in \mathbb{C}| \sup_{\theta \in [-\frac{\pi}{2}, \frac{\pi}{2}]} Re(-ze^{i\theta}) < \pi\}$$

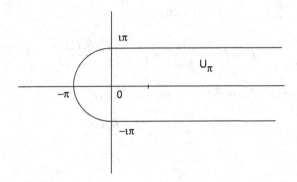

Then if $z \in U_\pi \setminus \{0\}$ we can write

$$\frac{e^{-zx}}{1 - e^{-z}} - \frac{e^{-z(x+1)}}{1 - e^{-z}} = e^{-zx}$$

this gives $R_f(x) = \frac{e^{-zx}}{1-e^{-z}} - \int_1^2 \frac{e^{-zx}}{1-e^{-z}} dx$, thus we have

$$R_f(x) = \frac{e^{-zx}}{1 - e^{-z}} - \frac{e^{-z}}{z}$$

and we get for $z \in U_\pi \setminus \{0\}$

$$\sum_{n \geq 1}^{\mathcal{R}} e^{-zn} = \frac{e^{-z}}{1 - e^{-z}} - \frac{e^{-z}}{z} \tag{1.15}$$

For $z = 0$ we have $\sum_{n \geq 1}^{\mathcal{R}} e^{-zn} = \sum_{n \geq 1}^{\mathcal{R}} 1 = \frac{1}{2} = \lim_{z \to 0} (\frac{e^{-z}}{1-e^{-z}} - \frac{e^{-z}}{z})$.

If $z = it$ with $t \in]-\pi, \pi[\setminus \{0\}$, taking real and imaginary part of (1.15) we get

$$\sum_{n \geq 1}^{\mathcal{R}} \cos(nt) = \frac{\sin(t)}{t} - \frac{1}{2} \tag{1.16}$$

$$\sum_{n \geq 1}^{\mathcal{R}} \sin(nt) = \frac{1}{2} \cot(\frac{t}{2}) - \frac{\cos(t)}{t} \tag{1.17}$$

(2) Let k be a positive integer. By the definition of the Bernoulli polynomials we verify that

$$\frac{B_{k+1}(x+1)}{k+1} - \frac{B_{k+1}(x)}{k+1} = x^k$$

and since $\int_0^1 B_{k+1}(x)dx = 0$ we have

$$\int_1^2 \frac{B_{k+1}(x)}{k+1}dx = \int_0^1 \frac{B_{k+1}(x+1)}{k+1}dx = \int_0^1 \frac{B_k(x)}{k+1}dx + \int_0^1 x^k dx = \frac{1}{k+1}$$

Thus if $f(x) = x^k$ where k is an integer ≥ 0 then

$$R_{x^k}(x) = \frac{1 - B_{k+1}(x)}{k+1} \tag{1.18}$$

thus we get

$$\sum_{n\geq 1}^{\mathcal{R}} n^k = \frac{1 - B_{k+1}}{k+1} \text{ if } k \geq 1 \text{ and } \sum_{n\geq 1}^{\mathcal{R}} 1 = \frac{1}{2} \tag{1.19}$$

(3) Take $f(x) = \frac{1}{x^z}$ with $Re(x) > 0$.

(a) For $Re(z) > 1$ and $Re(x) > 0$ the series $\sum_{n\geq 0} \frac{1}{(n+x)^z}$ is convergent and defines the *Hurwitz zeta function:*

$$\zeta(z,x) = \sum_{n=0}^{+\infty} \frac{1}{(n+x)^z}$$

Since

$$\zeta(z,x) - \zeta(z,x+1) = \frac{1}{x^z} \tag{1.20}$$

we have

$$R_f(x) = \zeta(z,x) - \int_1^2 \sum_{n=0}^{+\infty} \frac{1}{(n+x)^z} dx$$

Since the series $\sum_{n=0}^{+\infty} \frac{1}{(n+x)^z}$ is uniformly convergent for $x \in [1,2]$ we have

$$\int_1^2 \sum_{n=0}^{+\infty} \frac{1}{(n+x)^z} dx = \frac{1}{z-1} \sum_{n=0}^{+\infty} \left(\frac{1}{(n+1)^{z-1}} - \frac{1}{(n+2)^{z-1}}\right) = \frac{1}{z-1}$$

Thus for $Re(z) > 1$ we have

$$R_{\frac{1}{x}}(x) = \zeta(z,x) - \frac{1}{z-1}$$

(b) For every $z \neq 1$ the integral formula (1.13) for the function R_f gives

$$R_{\frac{1}{x}}(x) = \frac{x^{1-z}}{z-1} - \frac{1}{z-1} + \frac{1}{2}x^{-z} + 2\int_0^{+\infty} \frac{(x^2+t^2)^{-z/2}\sin(z\arctan(t/x))}{e^{2\pi t}-1}dt$$

since this last integral defines an entire function of z we see that the Hurwitz zeta function, previously defined by the sum $\sum_{n=0}^{+\infty}\frac{1}{(n+x)^z}$, can be continued analytically for all $z \neq 1$ by

$$\zeta(z,x) = \frac{x^{1-z}}{z-1} + \frac{1}{2}x^{-z} + 2\int_0^{+\infty} \frac{(x^2+t^2)^{-z/2}\sin(z\arctan(t/x))}{e^{2\pi t}-1}dt$$

And by analytic continuation the Eq. (1.20) remains valid, thus for all $z \neq 1$ we have

$$R_{\frac{1}{x}}(x) = \zeta(z,x) - \frac{1}{z-1} \tag{1.21}$$

If we define for $z \neq 1$ the *Riemann zeta function* by

$$\zeta(z) = \zeta(z,1)$$

then we have

$$\sum_{n\geq 1}^{\mathcal{R}} \frac{1}{n^z} = \zeta(z) - \frac{1}{z-1} \tag{1.22}$$

We use the notation

$$\zeta^{\mathcal{R}}(z) = \sum_{n\geq 1}^{\mathcal{R}} \frac{1}{n^z} \tag{1.23}$$

(4) If $f(x) = \frac{1}{x}$ then the series $\sum_{n=0}^{+\infty}\frac{1}{n+x}$ is divergent and to get a solution of the difference equation $R(x) - R(x+1) = \frac{1}{x}$ we replace it by the series $\sum_{n=0}^{+\infty}(\frac{1}{n+x} - \frac{1}{n+1})$. Thus we have

$$R_f(x) = \sum_{n=0}^{+\infty}(\frac{1}{n+x} - \frac{1}{n+1}) - \int_1^2 \sum_{n=0}^{+\infty}(\frac{1}{n+x} - \frac{1}{n+1})dx$$

This last integral is

$$\int_1^2 \sum_{n=0}^{+\infty} (\frac{1}{n+x} - \frac{1}{n+1})dx = \sum_{n=1}^{+\infty}(Log(n+1) - Log(n) - \frac{1}{n}) = -\gamma$$

where γ is the Euler constant. Thus

$$R_{\frac{1}{x}}(x) = \sum_{n=0}^{+\infty} (\frac{1}{n+x} - \frac{1}{n+1}) + \gamma$$

and we get

$$\sum_{n\geq 1}^{\mathcal{R}} \frac{1}{n} = \gamma \tag{1.24}$$

(5) Another way to get (1.24) is to use the the *digamma function* $\psi = \Gamma'/\Gamma$ that verifies

$$\psi(x+1) - \psi(x) = \frac{1}{x} \text{ with } \int_1^2 \psi(x)dx = 0$$

Thus we have

$$R_{\frac{1}{x}} = -\psi(x) \tag{1.25}$$

Note that we have

$$\gamma + \psi(x) = \sum_{n=0}^{+\infty}(\frac{1}{n+1} - \frac{1}{n+x}) \tag{1.26}$$

and $\psi(1) = -\gamma$. By derivation we have for any integer $k > 0$

$$\partial^k \psi(x+1) - \partial^k \psi(x) = (-1)^k \frac{k!}{x^{k+1}}$$

with

$$\int_1^2 \partial^k \psi(x)dx = \partial^{k-1}\psi(2) - \partial^{k-1}\psi(1) = (-1)^{k-1}(k-1)!$$

Thus we get

$$R_{\frac{1}{x^{k+1}}} = \frac{(-1)^{k-1}}{k!}\partial^k\psi - \frac{1}{k} \tag{1.27}$$

Note that for any integer $k > 0$ we have $\sum_{n\geq 1}^{\mathcal{R}} \frac{1}{n^{k+1}} = \zeta(k+1) - \frac{1}{k}$, thus we have

$$\frac{\partial^k \psi(1)}{k!} = (-1)^{k+1}\zeta(k+1)$$

(6) Take $f(x) = Log(x)$, where Log is the principal determination of the logarithm. Then the relation $\Gamma(x+1) = x\Gamma(x)$ gives

$$Log\ \Gamma(x+1) - Log\ \Gamma(x) = Log(x)$$

thus

$$R_{Log} = -Log\ \Gamma + \int_1^2 Log\ \Gamma(t)dt$$

since it is known that $\int_1^2 Log\ \Gamma(t)dt = -1 + Log(\sqrt{2\pi})$ cf. Srivastava and Choi) we get

$$R_{Log}(x) = -Log(\Gamma(x)) + Log(\sqrt{2\pi}) - 1 \qquad (1.28)$$

and for $x = 1$ we have

$$\sum_{n\geq 1}^{\mathcal{R}} Log(n) = Log(\sqrt{2\pi}) - 1$$

1.4.2 The Fractional Sum

We will now explain how the Ramanujan summation is related to the function that Ramanujan writes

$$\varphi(x) = f(1) + \ldots + f(x)$$

Consider $f \in \mathcal{O}^\pi$ and suppose we have a function φ analytic for $Re(x) > a$ with $-1 < a < 0$ and of exponential type $< \pi$ which satisfies

$$\varphi(x) - \varphi(x-1) = f(x) \text{ with } \varphi(0) = 0$$

This gives $\varphi(n) = f(1) + \ldots + f(n)$ for every positive integer n, thus the function φ is an interpolation function of the partial sums of the series $\sum f(n)$. If we set

$$R(x) = -\varphi(x-1) + \int_0^1 \varphi(x)dx$$

then R is an analytic function for $Re(x) > a + 1$ of exponential type $< \pi$, which satisfies $R(x) - R(x+1) = f(x)$ with $\int_1^2 R(x)dx = 0$, thus $R = R_f$ and we get

$$\varphi(x) = \int_0^1 \varphi(x)dx - R_f(x+1)$$

Since $\varphi(0) = 0$ we have

$$\int_0^1 \varphi(x)dx = R_f(1) = \sum_{n \geq 1}^{\mathcal{R}} f(n)$$

and the function φ is simply related to the function R_f by

$$\varphi(x) = R_f(1) - R_f(x+1)$$

Inversely this last equation defines a function φ analytic for $Re(x) > a$ with $-1 < a < 0$ and of exponential type $< \pi$ which satifies

$$\varphi(x) - \varphi(x-1) = f(x) \text{ with } \varphi(0) = 0$$

We are thus led to the following definition:

Definition 3 Let $f \in \mathcal{O}^\pi$, there is a unique function φ_f analytic for $Re(x) > a$ with $-1 < a < 0$ and of exponential type $< \pi$ which satisfies

$$\varphi_f(x) - \varphi_f(x-1) = f(x) \text{ with } \varphi_f(0) = 0$$

We call this function the *fractional sum of f*, it is related to R_f by

$$\varphi_f(x) = R_f(1) - R_f(x+1) = \sum_{n \geq 1}^{\mathcal{R}} f(n) - R_f(x) + f(x) \tag{1.29}$$

and we have

$$\sum_{n \geq 1}^{\mathcal{R}} f(n) = \int_0^1 \varphi_f(x)dx \tag{1.30}$$

Example If k is an integer ≥ 1 then

$$\varphi_{x^k}(x) = \frac{1 - B_{k+1}}{k+1} - \frac{1 - B_{k+1}(x)}{k+1} + x^k = \frac{B_{k+1}(x) - B_{k+1}}{k+1} + x^k$$

If $k = 0$ we have $\varphi_1(x) = B_1(x) + \frac{1}{2} = x$.

If k is an integer ≥ 1 then

$$\varphi_{\frac{1}{x^{k+1}}} = \frac{(-1)^k}{k!}\partial^k\Psi + \frac{1}{x^{k+1}} + \zeta(k+1)$$

For $k = 0$ we have

$$\varphi_{\frac{1}{x}}(x) = \Psi(x) + \frac{1}{x} + \gamma$$

More generally for $z \neq 1$ we have

$$\varphi_{\frac{1}{x^z}}(x) = \sum_{n\geq 1}^{\mathcal{R}} \frac{1}{n^z} - (\zeta(z,x) - \frac{1}{z-1}) + \frac{1}{x^z} = \zeta(z) - \zeta(z,x) + \frac{1}{x^z}$$

Integral Expression of φ_f

The relation (1.29) between φ_f and R_f gives by (1.13)

$$\varphi_f(x) = \sum_{n\geq 1}^{\mathcal{R}} f(n) + \frac{f(x)}{2} + \int_1^x f(t)dt - i\int_0^{+\infty} \frac{f(x+it)-f(x-it)}{e^{2\pi t}-1}dt$$

If we replace x by a positive integer n we get the classic *Abel-Plana formula*

$$f(1) + \ldots + f(n) = \frac{f(1)}{2} + i\int_0^{+\infty} \frac{f(1+it)-f(1-it)}{e^{2\pi t}-1}dt$$
$$+\frac{f(n)}{2} + \int_1^n f(t)dt - i\int_0^{+\infty} \frac{f(n+it)-f(n-it)}{e^{2\pi t}-1}dt$$

Note that the condition

$$\lim_{n\to+\infty} \int_0^{+\infty} \frac{f(n+it)-f(n-it)}{e^{2\pi t}-1}dt = 0$$

is sufficient to obtain

$$\sum_{n\geq 1}^{\mathcal{R}} f(n) = \lim_{n\to+\infty} \left[f(1) + \ldots + f(n) - \frac{f(n)}{2} - \int_1^n f(t)dt\right]$$

Examples

(1) Take the function $f = x \mapsto \frac{1}{p+xq}$ where p and q are strictly positive integers. We have

$$\sum_{n\geq 1}^{\mathcal{R}} \frac{1}{p+nq} = \lim_{n\to+\infty} \left(\frac{1}{p+q} + \ldots + \frac{1}{p+nq} - \frac{1}{q}Log(nq)\right) + \frac{1}{q}Log(p+q)$$

Thus the generalized Euler constant (Lehmer 1975) defined by

$$\gamma(p,q) = \lim_{n\to+\infty} \left(\frac{1}{p} + \frac{1}{p+q} + \ldots + \frac{1}{p+nq} - \frac{1}{q}Log(nq)\right)$$

is related to the Ramanujan summation by

$$\gamma(p,q) = \sum_{n\geq 1}^{\mathcal{R}} \frac{1}{p+nq} + \frac{1}{p} - \frac{1}{q}Log(p+q)$$

(2) Take $f(x) = Log(x)$ then

$$\varphi_{Log}(x) = Log(\Gamma(x+1))$$

and by the integral expression of φ_{Log} we have

$$Log(\Gamma(x+1)) = Log(\sqrt{2\pi}) + \frac{1}{2}Log(x) + xLog(x) - x + 2\int_0^{+\infty} \frac{\arctan(\frac{t}{x})}{e^{2\pi t}-1}dt$$

Remark In his Notebook Ramanujan uses for the function φ_f the expression

$$\varphi_f(x) = \sum_{n=1}^{+\infty}(f(n)-f(n+x)) \tag{1.31}$$

when this series is convergent.

Note that if the series $\sum_{n\geq 1}(f(n)-f(n+x))$ is convergent and if

$$\lim_{x\to+\infty} f(x) = 0$$

then the relation

$$\sum_{n=1}^{N}(f(n)-f(n+x)) - \sum_{n=1}^{N}(f(n)-f(n+x-1)) = f(x)-f(x+N)$$

gives when $N \to +\infty$

$$\varphi_f(x) - \varphi_f(x-1) = f(x)$$

Thus if we assume that the function defined by (1.31) is analytic for $Re(x) > a$ with $-1 < a < 0$ and of exponential type $< \pi$ then it verifies the requirements of Definition 3.

1.4.3 Relation to Usual Summation

Consider a function $f \in \mathcal{O}^\pi$, such that

$$\lim_{n \to +\infty} f(n) = 0$$

By the definition of R_f we have

$$R_f(1) - R_f(n) = \sum_{k=1}^{n-1} f(k).$$

Thus the series $\sum_{n \geq 1} f(n)$ is convergent if and only if $R_f(n)$ has a finite limit when $n \to +\infty$ and in this case

$$\sum_{n \geq 1}^{\mathcal{R}} f(n) = \sum_{n \geq 1}^{\infty} f(n) + \lim_{n \to +\infty} R_f(n)$$

Now let us see how this last limit is simply related to the integral of the function f. Recall the integral formula

$$R_f(n) = \frac{f(n)}{2} - \int_1^n f(t) dt + i \int_0^{+\infty} \frac{f(n+it) - f(n-it)}{e^{2\pi t} - 1} dt$$

and assume that

$$\lim_{n \to +\infty} \int_0^{+\infty} \frac{f(n+it) - f(n-it)}{e^{2\pi t} - 1} dt = 0$$

Then the convergence of the integral $\int_1^{+\infty} f(t) dt$ is equivalent to the fact that $R_f(n)$ has a finite limit when $n \to +\infty$ and in this case

$$\lim_{n \to +\infty} R_f(n) = - \int_1^{+\infty} f(t) dt$$

Thus we have proved the following result:

Theorem 2 *Consider $f \in \mathcal{O}^\pi$ such that*

$$\lim_{n \to +\infty} f(n) = 0$$

and

$$\lim_{n \to +\infty} \int_0^{+\infty} \frac{f(n+it) - f(n-it)}{e^{2\pi t} - 1} dt = 0 \qquad (1.32)$$

Then the series $\sum_{n \geq 1} f(n)$ is convergent if and only if the integral $\int_1^{+\infty} f(t)dt$ is convergent and we have

$$\sum_{n \geq 1}^{\mathcal{R}} f(n) = \sum_{n \geq 1}^{\infty} f(n) - \int_1^{+\infty} f(x)dx \qquad (1.33)$$

From now on, we will say in this case that "we are in a case of convergence".

Example For $z \in \mathbb{C} \setminus \{1, 2, 3, \ldots\}$ we have

$$\sum_{n \geq 1}^{\mathcal{R}} (\frac{1}{n} - \frac{1}{n+z}) = \sum_{n \geq 1}^{\infty} (\frac{1}{n} - \frac{1}{n+z}) - \int_1^{+\infty} (\frac{1}{x} - \frac{1}{x+z}) dx$$

and by (1.26) we get

$$\sum_{n \geq 1}^{\mathcal{R}} \frac{1}{n} - \sum_{n \geq 1}^{\mathcal{R}} \frac{1}{n+z} = \gamma + \psi(z+1) - \int_1^{+\infty} (\frac{1}{x} - \frac{1}{x+z}) dx$$

Thus for $z \in \mathbb{C} \setminus \{1, 2, 3, \ldots\}$ we have

$$\sum_{n \geq 1}^{\mathcal{R}} \frac{1}{n+z} = -\psi(z+1) + Log(z+1) \qquad (1.34)$$

If p and q are strictly positive integers this gives

$$\sum_{n \geq 1}^{\mathcal{R}} \frac{1}{p+nq} = -\frac{1}{q}\psi(\frac{p}{q}) - \frac{1}{p} + \frac{1}{q}Log(\frac{p}{q} + 1)$$

By Gauss formula for ψ (c.f. Srivastava and Choi) we get for $0 < p < q$

$$\sum_{n\geq1}^{\mathcal{R}} \frac{1}{p+nq} = \frac{1}{q}\gamma - \frac{1}{p} + \frac{1}{q}Log(p+q) + \frac{\pi}{2q}\cot(\pi\frac{p}{q})$$

$$- \frac{1}{q}\sum_{k=1}^{q-1}\cos(2\pi k\frac{p}{q})Log(2\sin(\pi\frac{k}{q}))$$

Remark Take a function $f \in \mathcal{O}^\pi$ such that the function

$$g : x \mapsto \sum_{n=0}^{+\infty}f(x+n)$$

is well defined for $Re(x) > 0$ and assume that $g \in \mathcal{O}^\pi$. Then we have

$$g(x) - g(x+1) = f(x)$$

Thus we deduce that

$$R_f(x) = g(x) - \int_1^2 g(x)dx$$

If in addition we have

$$\int_1^2 \sum_{n=0}^{+\infty}f(x+n)dx = \sum_{n=0}^{+\infty}\int_1^2 f(x+n)dx$$

then it follows that

$$R_f(x) = \sum_{n=0}^{+\infty}f(x+n) - \int_1^{+\infty}f(x)dx. \tag{1.35}$$

Thus we see that in this case the function R_f is simply related to the usual remainder of the convergent series $\sum_{n\geq1}f(n)$.

Chapter 2
Properties of the Ramanujan Summation

In this chapter we give some properties of the Ramanujan summation in comparison to the usual properties of the summation of convergent series.

In the first section we begin with the shift property which replaces the usual translation property of convergent series. This has important consequences in the use of the Ramanujan summation, especially for the classical formula of summation by parts and of summation of a product. General functional relations for the fractional sum φ_f are also deduced.

In the second section we examine the relation between the Ramanujan summation and derivation, we see that the fractional sums φ_f and $\varphi_{\partial f}$ are simply related. We deduce some simple formulas for the evaluation of the Ramanujan summation of some series, which constitutes the content of Theorems 5 and 6.

In the third section we show that in the case of the Ramanujan summation of an entire function the sum $\sum_{n\geq 1}^{\mathcal{R}} f(n)$ can be expanded in a convergent series involving the Bernoulli numbers. This very easily gives some formulas involving classical constant or trigonometric series.

2.1 Some Elementary Properties

2.1.1 The Unusual Property of the Shift

Let $f \in \mathcal{O}^\pi$ and consider the function g defined by $g(u) = f(u + 1)$, we have by definition of R_f

$$R_f(u + 1) - R_f(u + 2) = f(u + 1) = g(u)$$

© Springer International Publishing AG 2017
B. Candelpergher, *Ramanujan Summation of Divergent Series*,
Lecture Notes in Mathematics 2185, DOI 10.1007/978-3-319-63630-6_2

we deduce that the fractional remainder function R_g is given by

$$R_g(u) = R_f(u+1) - \int_1^2 R_f(u+1)du$$

$$= R_f(u) - f(u) - \int_1^2 (R_f(u) - f(u))du$$

Thus we have

$$R_g(u) = R_f(u) - f(u) + \int_1^2 f(u)du$$

And for $u = 1$ this gives the *shift property*

$$\sum_{n \geq 1}^{\mathcal{R}} f(n+1) = \sum_{n \geq 1}^{\mathcal{R}} f(n) - f(1) + \int_1^2 f(x)dx \qquad (2.1)$$

Therefore we see that the Ramanujan summation *does not satisfy* the usual translation property of convergent series.

More generally let $f \in \mathcal{O}^\pi$ and for $\mathrm{Re}(x) \geq 0$ let $g(u) = f(u+x)$. Then we have

$$R_f(u+x) - R_f(u+x+1) = f(u+x) = g(u)$$

we deduce that

$$R_g(u) = R_f(u+x) - \int_{x+1}^{x+2} R_f(v)dv$$

thus taking $u = 1$ in this equation we get

$$\sum_{n \geq 1}^{\mathcal{R}} f(n+x) = R_f(x+1) - \int_{x+1}^{x+2} R_f(v)dv$$

This last integral can be evaluated by using

$$\int_1^{x+1} R_f(v)dv - \int_1^{x+1} R_f(v+1)dv = \int_1^{x+1} f(v)dv$$

which gives

$$-\int_{x+1}^{x+2} R_f(v)dv = \int_1^{x+1} f(v)dv$$

Therefore we have

$$\sum_{n\geq1}^{\mathcal{R}} f(n+x) = R_f(x+1) + \int_1^{x+1} f(v)dv = R_f(x) - f(x) + \int_1^{x+1} f(v)dv \qquad (2.2)$$

Since $R_f(x+1) = R_f(1) - \varphi_f(x)$ we get the *general shift property*

$$\sum_{n\geq1}^{\mathcal{R}} f(n+x) = \sum_{n\geq1}^{\mathcal{R}} f(n) - \varphi_f(x) + \int_1^{x+1} f(v)dv \qquad (2.3)$$

If m is a positive integer then the shift property is simply

$$\sum_{n\geq1}^{\mathcal{R}} f(n+m) = \sum_{n\geq1}^{\mathcal{R}} f(n) - \sum_{n=1}^{m} f(n) + \int_1^{m+1} f(x)dx$$

Remark Let us consider a function f such that $x \mapsto f(x-1)$ is in $\mathcal{O}^\pi$, we define $\sum_{n\geq0}^{\mathcal{R}} f(n)$ by

$$\sum_{n\geq0}^{\mathcal{R}} f(n) = \sum_{n\geq1}^{\mathcal{R}} f(n-1)$$

By the shift property applied to $x \mapsto f(x-1)$ we have

$$\sum_{n\geq0}^{\mathcal{R}} f(n) = f(0) + \sum_{n\geq p}^{\mathcal{R}} f(n) + \int_0^1 f(x)dx$$

Examples

(1) Let $f(x) = \frac{1}{x}$ and $H_m = \sum_{k=1}^{m} \frac{1}{k}$, we have

$$\sum_{n\geq1}^{\mathcal{R}} \frac{1}{n+m} = \sum_{n\geq1}^{\mathcal{R}} \frac{1}{n} - H_m + \int_1^{m+1} \frac{1}{x}dx = \gamma - H_m + Log(m+1)$$

(which is a special case of (1.34) since $H_m = \psi(m+1) + \gamma$).
   As an application, consider a rational function

$$g(x) = \sum_{m=0}^{M} \frac{c_m}{x+m} \quad \text{such that} \quad \sum_{m=0}^{M} c_m = 0$$

then we are in a case of convergence for the series $\sum_{n \geq 1} g(n)$ and we have

$$\sum_{n=1}^{+\infty} g(n) = \sum_{n \geq 1}^{\mathcal{R}} \sum_{m=0}^{M} \frac{c_m}{n+m} + \int_1^{+\infty} \sum_{m=0}^{M} \frac{c_m}{x+m} dx$$

Note that

$$\sum_{n \geq 1}^{\mathcal{R}} \sum_{m=0}^{M} \frac{c_m}{n+m} = \sum_{m=0}^{M} \sum_{n \geq 1}^{\mathcal{R}} \frac{c_m}{n+m} = -\sum_{m=1}^{M} c_m H_m + \sum_{m=1}^{M} c_m Log(m+1)$$

and

$$\int_1^{+\infty} \sum_{m=0}^{M} \frac{c_m}{x+m} dx = \lim_{A \to +\infty} \int_1^A \sum_{m=0}^{M} \frac{c_m}{x+m} dx = -\sum_{m=1}^{M} c_m Log(1+m)$$

Thus we get

$$\sum_{n=1}^{+\infty} g(n) = -\sum_{m=1}^{M} c_m H_m$$

By summation by parts we obtain the classical result

$$\sum_{n=1}^{+\infty} g(n) = c_0 + \frac{c_0 + c_1}{2} + \ldots + \frac{c_0 + \ldots + c_{m-1}}{m}$$

(2) For $f(x) = Log(x)$ we get by (2.2)

$$\sum_{n \geq 1}^{\mathcal{R}} Log(n+x) = -Log(\Gamma(x+1)) + Log(\sqrt{2\pi}) - 1 + (x+1)Log(x+1) - x$$

Let's have $t > 0$, since

$$\arctan(\frac{t}{n}) = \frac{1}{2i}(Log(n+it) - Log(n-it))$$

then we get

$$\sum_{n \geq 1}^{\mathcal{R}} \arctan(\frac{t}{n}) = \frac{1}{2i}(\sum_{n \geq 1}^{\mathcal{R}} Log(n+it) - \sum_{n \geq 1}^{\mathcal{R}} Log(n-it))$$

$$= -\frac{1}{2i} Log(\frac{\Gamma(1+it)}{\Gamma(1-it)}) + tLog(\sqrt{t^2+1}) + \arctan(t) - t$$

With $t = 1$ we get

$$\sum_{n \geq 1}^{\mathcal{R}} \arctan(n) = \sum_{n \geq 1}^{\mathcal{R}} \frac{\pi}{2} - \sum_{n \geq 1}^{\mathcal{R}} \arctan(\frac{1}{n})$$

$$= -\frac{1}{2i} Log(\frac{\Gamma(1+i)}{\Gamma(1-i)}) + Log(\sqrt{2}) + \frac{\pi}{2} - 1$$

Note also that we have $\arctan(\frac{t}{n}) - \frac{t}{n} = O(\frac{1}{n^2})$, thus we are in a case of convergence and by (1.33) we deduce that

$$\sum_{n=1}^{+\infty} (\arctan(\frac{t}{n}) - \frac{t}{n}) = \sum_{n \geq 1}^{\mathcal{R}} \arctan(\frac{t}{n}) - \gamma t + \int_1^{+\infty} (\arctan(\frac{t}{x}) - \frac{t}{x}) dx$$

this last integral is $\int_0^t (\arctan(u) - u) \frac{1}{u^2} du$ and by integration by parts we get

$$\sum_{n=1}^{+\infty} (\arctan(\frac{t}{n}) - \frac{t}{n}) = -\frac{1}{2i} Log(\frac{\Gamma(1+it)}{\Gamma(1-it)}) - \gamma t$$

(3) Let $t \in]-\pi, +\pi[$ and consider the series

$$\sum_{n \geq 1} (\frac{1}{1 + e^{(n-1)t}} - \frac{1}{1 + e^{nt}})$$

The partial sum of order N of this series is equal to $\frac{1}{2} - \frac{1}{1 + e^{Nt}}$, this gives the convergence of the series and we have

$$\sum_{n=1}^{+\infty} (\frac{1}{1 + e^{(n-1)t}} - \frac{1}{1 + e^{nt}}) = \begin{cases} -\frac{1}{2} & \text{for } t < 0 \\ 0 & \text{for } t = 0 \\ +\frac{1}{2} & \text{for } t > 0 \end{cases}$$

Thus, although the terms of the series are continuous at $t = 0$, the usual sum has a discontinuity for $t = 0$. *This is not the case for the Ramanujan summation,* as we can verify with this example:

By the shift property applied to the function

$$f(x) = \frac{1}{1 + e^{(x-1)t}}$$

we have immediately

$$\sum_{n \geq 1}^{\mathcal{R}} (\frac{1}{1 + e^{(n-1)t}} - \frac{1}{1 + e^{nt}}) = \frac{1}{2} - \int_0^1 \frac{1}{1 + e^{xt}} dx$$

This is a continuous function of t which is given by

$$\sum_{n\geq 1}^{\mathcal{R}}\left(\frac{1}{1+e^{(n-1)t}}-\frac{1}{1+e^{nt}}\right) = \begin{cases} -\frac{1}{2}+\frac{1}{t}(Log(1+e^t)-Log(2)) & \text{for } t\neq 0 \\ 0 & \text{for } t=0 \end{cases}$$

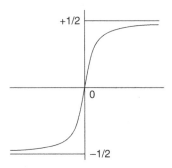

We examine in Chap. 3 how the Ramanujan summation of a series $\sum_{n\geq 1}f(n,z)$ with $f(x,z)$ depending analytically on an external complex parameter z conserves the property of analyticity.

Application

Ramanujan gives an explicit evaluation (Ramanujan 1927, 13. Sommation of a certain series) of sums like

$$\sum_{n=0}^{+\infty}(\sqrt{n+1}-\sqrt{n})^3 = \frac{3}{2\pi}\zeta(\frac{3}{2})$$

Let us see how we can obtain this result by a simple application of the shift property.
First we write

$$\sum_{n=0}^{+\infty}(\sqrt{n+1}-\sqrt{n})^3 = -1 + \sum_{n=1}^{+\infty}(\sqrt{n+1}-\sqrt{n})^3$$

and since we are in a case of convergence

$$\sum_{n=1}^{+\infty}(\sqrt{n+1}-\sqrt{n})^3 = \sum_{n\geq 1}^{\mathcal{R}}(\sqrt{n+1}-\sqrt{n})^3 + \int_{1}^{+\infty}(\sqrt{x+1}-\sqrt{x})^3 dx$$

$$= \sum_{n\geq 1}^{\mathcal{R}}(\sqrt{n+1}-\sqrt{n})^3 - \frac{1}{5}(12\sqrt{2}-18)$$

Next we use the binomial expansion of $(\sqrt{n+1} - \sqrt{n})^3$ and we get

$$\sum_{n\geq 1}^{\mathcal{R}}(\sqrt{n+1} - \sqrt{n})^3 = 4\sum_{n\geq 1}^{\mathcal{R}}((n+1)^{3/2} - n^{3/2}) - 3\sum_{n=1}^{\mathcal{R}}(n^{1/2} + (n+1)^{1/2})$$

By the shift property (2.1) we have

$$\sum_{n\geq 1}^{\mathcal{R}}((n+1)^{3/2} - n^{3/2}) = -1 + \int_1^2 x^{3/2}dx = -1 + \frac{1}{5}(8\sqrt{2} - 2)$$

$$\sum_{n\geq 1}^{\mathcal{R}}((n+1)^{1/2} + n^{1/2}) = 2\sum_{n\geq 1}^{\mathcal{R}}n^{1/2} - 1 + \int_1^2 x^{1/2}dx$$

$$= 2(\zeta(-\frac{1}{2}) + \frac{2}{3}) - 1 + \frac{1}{3}(4\sqrt{2} - 2)$$

After some simplifications we get

$$\sum_{n=0}^{+\infty}(\sqrt{n+1} - \sqrt{n})^3 = -6\zeta(-\frac{1}{2})$$

which, by the functional equation of the zeta function, is Ramanujan's result.

Note that the series $\sum_{n\geq 1}(\sqrt{n+1} - \sqrt{n})$ is divergent but we can evaluate the sum of the "regularized series" $\sum_{n\geq 1}(\sqrt{n+1} - \sqrt{n} - \frac{1}{2\sqrt{n}})$ by the same technique:

$$\sum_{n=1}^{+\infty}(\sqrt{n+1} - \sqrt{n} - \frac{1}{2\sqrt{n}}) = \sum_{n\geq 1}^{\mathcal{R}}\sqrt{n+1} - \sum_{n\geq 1}^{\mathcal{R}}\sqrt{n} - \sum_{n\geq 1}^{\mathcal{R}}\frac{1}{2\sqrt{n}}$$

$$+ \int_1^{+\infty}(\sqrt{x+1} - \sqrt{x} - \frac{1}{2\sqrt{x}})dx$$

$$= -\frac{1}{2}\zeta(\frac{1}{2}) - 1$$

2.1.2 Summation by Parts

Let $f \in \mathcal{O}^\pi$ and $g \in \mathcal{O}^\pi$ be two functions such that $fg \in \mathcal{O}^\pi$, we have

$$f(x)g(x) - f(x+1)g(x+1) = (f(x) - f(x+1))g(x) + f(x+1)(g(x) - g(x+1))$$

Thus by definition of the Ramanujan summation we have

$$\sum_{n\geq1}^{\mathcal{R}}(f(n)-f(n+1))g(n)+f(n+1)(g(n)-g(n+1))=f(1)g(1)-\int_1^2 f(t)g(t)dt$$

Using the property of linearity we get the *formula of summation by parts*

$$\sum_{n\geq1}^{\mathcal{R}}(f(n+1)-f(n))g(n)=-\sum_{n\geq1}^{\mathcal{R}}f(n+1)(g(n+1)-g(n))$$

$$-f(1)g(1)+\int_1^2 f(t)g(t)dt$$

(2.4)

Examples

(1) With $f(x)=Log(x)$ and $g(x)=x$ we have

$$\sum_{n\geq1}^{\mathcal{R}}n\,Log(1+\frac{1}{n})=-\sum_{n\geq1}^{\mathcal{R}}Log(n)+\int_0^1 t\,Log(t+1)dt$$

and we get

$$\sum_{n\geq1}^{\mathcal{R}}n\,Log(1+\frac{1}{n})=\frac{5}{4}-Log(\sqrt{2\pi})$$

The regularized series $\sum_{n\geq1}(n\,Log(1+\frac{1}{n})-1+\frac{1}{2n})$ is convergent and by linearity and application of (1.33) we deduce that

$$\sum_{n\geq1}^{+\infty}(n\,Log(1+\frac{1}{n})-1+\frac{1}{2n})=\frac{1}{2}\gamma-Log(\sqrt{2\pi})+1$$

2.1.3 Sums of Products

If we replace g by φ_g in the formula (2.4) of summation by parts we get

$$\sum_{n\geq1}^{\mathcal{R}}(f(n+1)-f(n))\varphi_g(n)=-\sum_{n\geq1}^{\mathcal{R}}f(n+1)(\varphi_g(n+1)-\varphi_g(n))$$

$$-f(1)\varphi_g(1)+\int_1^2 f(t)\varphi_g(t)dt$$

since $\varphi_g(n+1) - \varphi_g(n) = g(n+1)$ we get, by the shift property, a formula for the *summation of a product*:

$$\sum_{n\geq 1}^{\mathcal{R}} f(n)g(n) = \int_1^2 f(t)\varphi_g(t-1)dt - \sum_{n\geq 1}^{\mathcal{R}}(f(n+1)-f(n))\varphi_g(n) \qquad (2.5)$$

Example If $f(x) = x$ and $g(x) = Log(x)$ we get by (2.5)

$$\sum_{n\geq 1}^{\mathcal{R}} nLog(n) = \int_1^2 t\varphi_{Log}(t-1)dt - \sum_{n\geq 1}^{\mathcal{R}} \varphi_{Log}(n)$$

Since $\varphi_{Log}(x) = Log(\Gamma(x+1))$ we have

$$\sum_{n\geq 1}^{\mathcal{R}} nLog(n) = \int_1^2 tLog(\Gamma(t))dt - \sum_{n\geq 1}^{\mathcal{R}} Log(n!)$$

Note that we can evaluate more directly the sum $\sum_{n\geq 1}^{\mathcal{R}} nLog(n)$ by using the relation (Srivastava and Junesang Choi 2012)

$$\int_1^{x+1} Log(\Gamma(t))dt - \int_1^x Log(\Gamma(t))dt = xLog(x) - x + Log(\sqrt{2\pi})$$

which gives

$$\sum_{n\geq 1}^{\mathcal{R}} nLog(n) - \frac{1-B_2}{2} + \frac{1}{2}Log(\sqrt{2\pi}) = \int_1^2 \int_1^x Log(\Gamma(t))dt\, dx$$

By integration by parts we deduce that

$$\sum_{n\geq 1}^{\mathcal{R}} nLog(n) = -\int_1^2 tLog(\Gamma(t))dt - \frac{19}{12} + \frac{3}{2}Log(\sqrt{2\pi})$$

Thus we get the relation

$$\sum_{n\geq 1}^{\mathcal{R}} Log(n!) = -2\sum_{n\geq 1}^{\mathcal{R}} nLog(n) - \frac{19}{12} + \frac{3}{2}Log(\sqrt{2\pi})$$

Theorem 3 *Let f and g be two functions of moderate growth then we have*

$$\sum_{n\geq 1}^{\mathcal{R}} f(n)g(n) = \sum_{n\geq 1}^{\mathcal{R}} \varphi_f(n)g(n) + \sum_{n\geq 1}^{\mathcal{R}} \varphi_g(n)f(n) - \sum_{n\geq 1}^{\mathcal{R}} f(n) \sum_{n\geq 1}^{\mathcal{R}} g(n)$$

$$- \int_1^2 R_f(x)R_g(x)dx$$

Proof Since $R_f(x) = \sum_{n\geq 1}^{\mathcal{R}} f(n) - \varphi_f(x) + f(x)$ we see that it is equivalent to prove that

$$\sum_{n\geq 1}^{\mathcal{R}} R_f(n)g(n) + \sum_{n\geq 1}^{\mathcal{R}} R_g(n)f(n) = \sum_{n\geq 1}^{\mathcal{R}} f(n)g(n) + \sum_{n\geq 1}^{\mathcal{R}} f(n) \sum_{n\geq 1}^{\mathcal{R}} g(n)$$

$$- \int_1^2 R_f(x)R_g(x)dx$$

We have immediately

$$R_f(x)R_g(x) - R_f(x+1)R_g(x+1) = R_f(x)g(x) + f(x)R_g(x) - f(x)g(x)$$

thus if $h(x) = R_f(x)g(x) + f(x)R_g(x) - f(x)g(x)$ then we have

$$R_h(x) = R_f(x)R_g(x) - \int_1^2 R_f(t)R_g(t)dt$$

taking the value for $x = 1$ we obtain the result.   □

Example With $f(x) = g(x) = \frac{1}{x}$ the preceding theorem gives

$$\sum_{n\geq 1}^{\mathcal{R}} \frac{H_n}{n} = \frac{1}{2}(\zeta(2) - 1 + \gamma^2) + \frac{1}{2}\int_1^2 (\psi(t))^2 dt \qquad (2.6)$$

Remark Since $R_f = \sum_{n\geq 1}^{\mathcal{R}} f(n) - \varphi_f + f$ we can express the result of Theorem 3 only in terms of the fractional sums:

$$\sum_{n\geq 1}^{\mathcal{R}} f(n)g(n) = \sum_{n\geq 1}^{\mathcal{R}} (f(n)\varphi_g(n) + \varphi_f(n)g(n))$$

$$+ \int_1^2 (f(x)\varphi_g(x) + \varphi_f(x)g(x))dx$$

$$- \int_1^2 (f(x)g(x) + \varphi_f(x)\varphi_g(x))dx$$

2.2 Summation and Derivation

Let us first remark that if $g \in \mathcal{O}^\alpha$ then we have $\partial g \in \mathcal{O}^\alpha$. This is a consequence of Cauchy integral formula, since if g is analytic for $Re(x) > a$ $(a < 1)$ then for $0 < r < 1 - a$ we have for $Re(x) > a + r$

$$|\partial g(x)| = |\frac{1}{2\pi r} \int_0^{2\pi} g(x + re^{i\theta})e^{-i\theta}d\theta| \le \frac{1}{2\pi r} \int_0^{2\pi} |g(x + re^{i\theta})|d\theta$$

Let $f \in \mathcal{O}^\pi$ then $\partial f \in \mathcal{O}^\pi$ and we have

$$\partial R_f(x) - \partial R_f(x + 1) = \partial f(x)$$

since $R_f \in \mathcal{O}^\pi$ then $\partial R_f \in \mathcal{O}^\pi$ and we deduce that

$$R_{\partial f} = \partial R_f - \int_1^2 \partial R_f(x)dx$$

This last integral is simply $R_f(2) - R_f(1) = -f(1)$ thus we get a relation between ∂R_f and $R_{\partial f}$ that is

$$R_{\partial f}(x) = \partial R_f(x) + f(1) \tag{2.7}$$

which we can translate into the fractional sums

$$\varphi_{\partial f}(x) = \partial \varphi_f(x) - f(1) + \sum_{n \ge 1}^{\mathcal{R}} \partial f(n) \tag{2.8}$$

Theorem 4 *Let $f \in \mathcal{O}^\pi$ then for every integer $m \ge 1$ we have*

$$\sum_{n \ge 1}^{\mathcal{R}} f(n) = -\sum_{k=1}^m \frac{B_k}{k!} \partial^{k-1}f(1) + (-1)^{m+1} \int_0^1 R_{\partial^m f}(t + 1) \frac{B_m(t)}{m!} dt \tag{2.9}$$

Proof We have $B_1(t) = t - \frac{1}{2}$ thus

$$0 = \int_0^1 R_f(t + 1)dt = \int_0^1 R_f(t + 1)\partial B_1(t)dt$$

Using the relation (2.7) and integrating by parts we obtain

$$R_f(1) = \frac{1}{2}f(1) + \int_0^1 R_{\partial f}(t + 1)B_1(t)dt$$

If we proceed by repeated integration by parts we find

$$\sum_{n\geq 1}^{\mathcal{R}} f(n) = -\sum_{k=1}^{m} \frac{B_k}{k!} \partial^{k-1} f(1) + (1)^{m+1} \int_0^1 R_{\partial^m f}(t+1)\frac{B_m(t)}{m!} dt$$

□

Corollary *Let $f \in \mathcal{O}^{\pi}$ and for $k \geq 1$ let*

$$F_k(x) = \int_1^x F_{k-1}(t)dt \text{ with } F_0 = f$$

then by the preceding theorem we have immediately

$$\sum_{n\geq 1}^{\mathcal{R}} F_m(n) = (-1)^{m+1} \int_0^1 R_f(t+1)\frac{B_m(t)}{m!} dt$$

For $m = 1$ we get

$$\sum_{n\geq 1}^{\mathcal{R}} \int_1^n f(x)dx = \int_1^2 t R_f(t)dt$$

Example Let $f(x) = \frac{1}{x}$. We have $F_1(x) = Log(x)$, and

$$F_k(x) = \frac{x^{k-1}}{(k-1)!}Log(x) + \frac{P_k(x)}{k!}$$

where the P_k are the polynomials defined by $P_1 = 0$ and

$$P_k'(x) = kP_{k-1}(x) - kx^{k-2} \text{ if } k \geq 2$$
$$P_k(1) = 0$$

By the preceding corollary we get

$$\sum_{n\geq 1}^{\mathcal{R}} n^k Log(n) = (-1)^{k-1} \int_0^1 \psi(t+1)\frac{B_{k+1}(t)}{k+1} dt - \sum_{n\geq 1}^{\mathcal{R}} \frac{P_{k+1}(n)}{k+1}$$

In this relation the last sum can be evaluated in terms of Bernoulli numbers since

$$\sum_{n\geq 1}^{\mathcal{R}} n^k = \frac{1 - B_{k+1}}{k+1} \text{ if } k \geq 1 \text{ and } \sum_{n\geq 1}^{\mathcal{R}} 1 = \frac{1}{2}$$

we get for example

$$\sum_{n\geq 1}^{\mathcal{R}} nLog(n) = \int_0^1 \psi(t+1)\frac{B_2(t)}{2!}dt - \frac{B_2}{2}$$

$$\sum_{n\geq 1}^{\mathcal{R}} n^2 Log(n) = -\int_0^1 \psi(t+1)\frac{B_3(t)}{3}dt + B_2 - \frac{1}{4}$$

Theorem 5 *Let $f \in \mathcal{O}^\pi$ then*

$$\sum_{n\geq 1}^{\mathcal{R}} \varphi_f(n) = \frac{3}{2}\sum_{n\geq 1}^{\mathcal{R}} f(n) - \sum_{n\geq 1}^{\mathcal{R}} nf(n) - \sum_{n\geq 1}^{\mathcal{R}} F(n) \text{ with } F(x) = \int_1^x f(t)dt$$

Proof Since we have

$$xR_f(x) - (x+1)R_f(x+1) = xf(x) - R_f(x+1)$$

we deduce that

$$\sum_{n\geq 1}^{\mathcal{R}} (nf(n) - R_f(n+1)) = R_f(1) - \int_1^2 tR_f(t)dt$$

By the preceding corollary we have $\int_1^2 tR_f(t)dt = \sum_{n\geq 1}^{\mathcal{R}} F(n)$, thus we obtain

$$\sum_{n\geq 1}^{\mathcal{R}} nf(n) - \sum_{n\geq 1}^{\mathcal{R}} R_f(n+1) = \sum_{n\geq 1}^{\mathcal{R}} f(n) - \sum_{n\geq 1}^{\mathcal{R}} F(n)$$

which is the result since $R_f(n+1) = R_f(1) - \varphi_f(n)$. $\qquad\square$

Example For all $s \in \mathbb{C}$ we define the harmonic numbers $H_n^{(s)}$ by

$$H_n^{(s)} = \varphi_{\frac{1}{x^s}}(n) = \sum_{k=1}^n \frac{1}{k^s}$$

Then by the preceding theorem we get

$$\sum_{n\geq 1}^{\mathcal{R}} H_n^{(s)} = \frac{3}{2}\sum_{n\geq 1}^{\mathcal{R}} \frac{1}{n^s} - \sum_{n\geq 1}^{\mathcal{R}} \frac{1}{n^{s-1}} - \sum_{n\geq 1}^{\mathcal{R}} \int_1^n t^{-s}dt$$

Let us write explicitly some different cases:

(a) For $s \neq 1, 2$ we have

$$\sum_{n \geq 1}^{\mathcal{R}} H_n^{(s)} = \frac{3}{2}\zeta(s) - \frac{3}{2}\frac{1}{s-1} - \zeta(s-1) + \frac{1}{s-2} - \frac{1}{s-1}\sum_{n \geq 1}^{\mathcal{R}}(1 - \frac{1}{n^{s-1}})$$

then

$$\sum_{n \geq 1}^{\mathcal{R}} H_n^{(s)} = \frac{3}{2}\zeta(s) - \frac{s-2}{s-1}\zeta(s-1) - \frac{1}{s-1}$$

(b) For $s = 1$ we have

$$\sum_{n \geq 1}^{\mathcal{R}} H_n = \frac{3}{2}\sum_{n \geq 1}^{\mathcal{R}}\frac{1}{n} - \sum_{n \geq 1}^{\mathcal{R}}1 - \sum_{n \geq 1}^{\mathcal{R}}Log(n)$$

thus

$$\sum_{n \geq 1}^{\mathcal{R}} H_n = \frac{3}{2}\gamma + \frac{1}{2} - Log(\sqrt{2\pi})$$

This gives the sum of the regularized series

$$\sum_{n=1}^{+\infty}(H_n - Log(n) - \gamma - \frac{1}{2n}) = \frac{1}{2}\gamma - Log(\sqrt{2\pi}) + \frac{1}{2}$$

(c) For $s = 2$ we have

$$\sum_{n \geq 1}^{\mathcal{R}} H_n^{(2)} = \frac{3}{2}\sum_{n \geq 1}^{\mathcal{R}}\frac{1}{n^2} - \sum_{n \geq 1}^{\mathcal{R}}\frac{1}{n} + \sum_{n \geq 1}^{\mathcal{R}}(\frac{1}{n} - 1)$$

thus

$$\sum_{n \geq 1}^{\mathcal{R}} H_n^{(2)} = \frac{3}{2}\zeta(2) - 2$$

This gives the sum of the regularized series

$$\sum_{n=1}^{+\infty}(H_n^{(2)} - \zeta(2) + \frac{1}{n}) = \zeta(2) - 1$$

Remark Let f be a function of moderate growth, then by Theorem 3 we have

$$\sum_{n\geq 1}^{\mathcal{R}} f(n)\partial f(n) = \sum_{n\geq 1}^{\mathcal{R}} \varphi_f(n)\partial f(n) + \sum_{n\geq 1}^{\mathcal{R}} \varphi_{\partial f}(n)f(n) - \sum_{n\geq 1}^{\mathcal{R}} f(n)\sum_{n\geq 1}^{\mathcal{R}} \partial f(n)$$
$$- \int_1^2 R_f(x)R_{\partial f}(x)dx$$

Using (2.7) this last integral term can be easily evaluated:

$$\int_1^2 R_f(x)R_{\partial f}(x)dx = \int_1^2 R_f(x)\partial R_f(x)dx = \frac{1}{2}f(1)^2 - f(1)R_f(1)$$

thus we find the relation

$$\sum_{n\geq 1}^{\mathcal{R}} f(n)\partial f(n) = \sum_{n\geq 1}^{\mathcal{R}} \varphi_f(n)\partial f(n) + \sum_{n\geq 1}^{\mathcal{R}} \varphi_{\partial f}(n)f(n) - \sum_{n\geq 1}^{\mathcal{R}} f(n)\sum_{n\geq 1}^{\mathcal{R}} \partial f(n)$$
$$- \frac{1}{2}f(1)^2 + f(1)\sum_{n\geq 1}^{\mathcal{R}} f(n)$$

For example if $f(x) = \frac{1}{x^p}$ with $p \neq 0, 1$ the preceding relation gives

$$\sum_{n\geq 1}^{\mathcal{R}} \frac{H_n^{(p)}}{n^{p+1}} + \sum_{n\geq 1}^{\mathcal{R}} \frac{H_n^{(p+1)}}{n^p} = \zeta(2p+1) + (\zeta(p) - \frac{1}{p-1})\zeta(p+1) - \frac{1}{p} \qquad (2.10)$$

and for $f(x) = \frac{1}{x}$ we obtain

$$\sum_{n\geq 1}^{\mathcal{R}} \frac{H_n}{n^2} + \sum_{n\geq 1}^{\mathcal{R}} \frac{H_n^{(2)}}{n} = \zeta(3) + \gamma\zeta(2) - 1 \qquad (2.11)$$

2.3 The Case of an Entire Function

2.3.1 The Sum of an Entire Function

In the first entry of Chapter XV of the second notebook Ramanujan writes:

$$h\varphi(h) + h\varphi(2h) + h\varphi(3h) + \ldots = \int_0^{+\infty} \varphi(x)dx + F(h)$$

where $F(h)$ can be found by expanding the left and writing the constant instead of a series and $F(0) = 0$.

   Cor. If $h\varphi(h) = ah^p + bh^q + ch^r + dh^s + \dots$ then

$$h\varphi(h) + h\varphi(2h) + \dots = \int_0^{+\infty} \varphi(x)dx + a\frac{B_p}{p}h^p\cos(\pi p/2) + b\frac{B_q}{q}h^q\cos(\pi q/2) + \dots$$

We try to give a precise meaning to Ramanujan's assertion with the following theorem.

Theorem 6 *Let f be an entire function of exponential type $< \pi$, then we have*

$$\sum_{n \geq 1}^{\mathcal{R}} f(n) = \int_0^1 f(x)dx - \frac{1}{2}f(0) - \sum_{k=1}^{+\infty} \partial^k f(0)\frac{B_{k+1}}{(k+1)!}$$

Thus in a case of convergence we have

$$\sum_{n \geq 1}^{+\infty} f(n) = \int_0^{+\infty} f(x)dx - \frac{1}{2}f(0) - \sum_{k=1}^{+\infty} \partial^k f(0)\frac{B_{k+1}}{(k+1)!}$$

Proof Let us write $f(x) = \sum_{k=0}^{+\infty} \frac{c_k}{k!}x^k$ with $c_k = \partial^k f(0)$. By the Cauchy integral formulas we have for $\beta < \pi$ a constant $C > 0$ such that for every integer $k \geq 0$ and every $r > 0$

$$|c_k| = k!r^{-k}\frac{1}{2\pi}|\int_0^{2\pi} f(re^{it})e^{-ikt}dt| \leq Cr^{-k}e^{\beta r}$$

Since $\beta < \pi$ we get

$$|c_k| \leq \inf_{r>0} Cr^{-k}e^{\beta r} \leq C\tau^k \text{ where } \tau < \pi$$

Let us now prove that $R_f = \sum_{k=0}^{+\infty} \frac{c_k}{k!}R_{x^k}$. We have

$$R_{x^k}(x) = \frac{1 - B_{k+1}(x)}{k+1} \text{ where } \frac{ze^{xz}}{e^z - 1} = \sum_{n \geq 0} \frac{B_n(x)}{n!}z^n$$

thus we consider the function

$$x \mapsto \sum_{k=0}^{+\infty} \frac{c_k}{k!}R_{x^k}(x) = \sum_{k=0}^{+\infty} \frac{c_k}{(k+1)!} - \sum_{k=0}^{+\infty} c_k\frac{B_{k+1}(x)}{(k+1)!}$$

By the Cauchy integral formula we have for $0 < r < 2\pi$

$$\frac{B_{k+1}(x)}{(k+1)!} = r^{-k}\frac{1}{2\pi}\int_0^{2\pi}\frac{e^{xre^{it}}}{e^{re^{it}}-1}e^{-ikt}dt$$

thus for $Re(x) > 0$ we get

$$\left|\frac{B_{k+1}(x)}{(k+1)!}\right| \leq \left(\frac{1}{2\pi}\int_0^{2\pi}\frac{1}{|e^{re^{it}}-1|}dt\right)r^{-k}e^{r|x|} = M_r r^{-k}e^{r|x|}$$

Since $|c_k| \leq C\tau^k$ where $\tau < \pi$, if we take $\tau < r < 2\pi$, this inequality shows that the series $\sum_{k=0}^{+\infty}c_k\frac{1-B_{k+1}(x)}{(k+1)!}$ is uniformly convergent on every compact of the x plane and defines an entire function of exponential type $< 2\pi$. By

$$\int_1^2\sum_{k=0}^{+\infty}c_k\frac{1-B_{k+1}(x)}{(k+1)!}dx = \sum_{k=0}^{+\infty}c_k\int_1^2\frac{1-B_{k+1}(x)}{(k+1)!}dx = 0$$

we can conclude that

$$R_f(x) = \sum_{k=0}^{+\infty}c_k\frac{1-B_{k+1}(x)}{(k+1)!}$$

Thus

$$R_f(1) = \sum_{k=0}^{+\infty}c_k\frac{1-B_{k+1}(1)}{(k+1)!} = \sum_{k=0}^{+\infty}\frac{c_k}{(k+1)!} - c_0 B_1(1) - \sum_{k=1}^{+\infty}c_k\frac{B_{k+1}(1)}{(k+1)!}$$

this gives

$$\sum_{n\geq1}^{\mathcal{R}}f(n) = \int_0^1 f(x)dx - \frac{1}{2}c_0 - \sum_{k=1}^{+\infty}c_k\frac{B_{k+1}}{(k+1)!}$$

$\square$

Remark Since $B_{2k+1} = 0$ for $k \geq 1$ we have

$$\sum_{n\geq1}^{\mathcal{R}}f(n) = \int_0^1 f(x)dx - \frac{1}{2}f(0) - \sum_{k=1}^{+\infty}\partial^{2k-1}f(0)\frac{B_{2k}}{(2k)!}$$

thus if f is even then

$$\sum_{n\geq 1}^{\mathcal{R}} f(n) = \int_0^1 f(x)dx - \frac{1}{2}f(0)$$

Example With p an integer > 0 and $0 < t < \pi/p$, let $f(x) = \frac{\sin^p(xt)}{x^p}$ for $x \neq 0$ and $f(0) = t^p$. This function is entire and even, thus by the preceding theorem we get

$$\sum_{n\geq 1}^{\mathcal{R}} \frac{\sin^p(nt)}{n^p} = \int_0^1 \frac{\sin^p(xt)}{x^p}dx - \frac{1}{2}t^p$$

Since we are in a case of convergence then

$$\sum_{n=1}^{+\infty} \frac{\sin^p(nt)}{n^p} = \sum_{n\geq 1}^{\mathcal{R}} \frac{\sin^p(nt)}{n^p} + \int_1^{+\infty} \frac{\sin^p(xt)}{x^p}dx = \int_0^{+\infty} \frac{\sin^p(xt)}{x^p}dx - \frac{1}{2}t^p$$

this gives

$$\sum_{n=1}^{+\infty} \frac{\sin^p(nt)}{n^p} = t^{p-1}\int_0^{+\infty} \frac{\sin^p(x)}{x^p}dx - \frac{1}{2}t^p$$

Remarks

(1) The preceding theorem is simply the relation

$$\sum_{n\geq 1}^{\mathcal{R}} f(n) = \sum_{k=0}^{+\infty} \frac{\partial^k f(0)}{k!} \sum_{n\geq 1}^{\mathcal{R}} n^k$$

If we apply this to the function

$$x \mapsto f(x+1) = \sum_{k=0}^{+\infty} \frac{\partial^k f(1)}{k!} x^k$$

we get

$$\sum_{n\geq 1}^{\mathcal{R}} f(n+1) = \sum_{k=0}^{+\infty} \frac{\partial^k f(1)}{k!} \sum_{n\geq 1}^{\mathcal{R}} n^k$$

and by the shift property we obtain

$$\sum_{n\geq 1}^{\mathcal{R}} f(n) = -\sum_{k=1}^{+\infty} \partial^{k-1} f(1) \frac{B_k}{k!}$$

(2) The preceding theorem is not valid if f is not an entire function. For example take $f(x) = \frac{1}{1+x^2 t^2}$, then if we apply the preceding result we find

$$\sum_{n\geq 1}^{+\infty} \frac{1}{1+n^2 t^2} = \int_0^{+\infty} \frac{1}{1+x^2 t^2} dx - \frac{1}{2} = \frac{\pi}{2t} - \frac{1}{2}$$

in contrast with the classical formula (Berndt 1985, Ramanujan's Notebooks II, ch.15, p.303)

$$\sum_{n\geq 1}^{+\infty} \frac{1}{1+n^2 t^2} = \frac{\pi}{2t} - \frac{1}{2} + \frac{\pi}{t} \frac{1}{e^{2\pi/t} - 1}$$

This is a particular case of a remark that Ramanujan writes after the first entry of chapter XV:

Il the expansion of $\varphi(h)$ be an infinite series then that of F(h) also will be an infinite series; but if most of the numbers p,q,r,s,t,... be odd integers F(h) appears to terminate. In this case the hidden part of F(h) can't be expanded in ascending powers of h and is very rapidly diminishing when h is slowly diminishing and consequently can be neglected for practical purposes when h is small.

Corollary *Let f be an entire function of exponential type $< \pi$. Then*

$$\sum_{n\geq 1}^{\mathcal{R}} \frac{f(n)}{n} = \int_0^1 \frac{f(x) - f(0)}{x} dx + \gamma f(0) - \frac{1}{2} f'(0) - \sum_{k=2}^{+\infty} \frac{\partial^k f(0)}{k!} \frac{B_k}{k}$$

Thus in a case of convergence we have

$$\sum_{n\geq 1}^{+\infty} \frac{f(n)}{n} = \int_0^1 \frac{f(x) - f(0)}{x} dx + \int_1^{+\infty} \frac{f(x)}{x} dx + \gamma f(0) - \frac{1}{2} f'(0)$$

$$- \sum_{k=2}^{+\infty} \frac{\partial^k f(0)}{k!} \frac{B_k}{k}$$

Proof Let the function g be defined by

$$g(x) = \frac{f(x) - f(0)}{x} \text{ if } x \neq 0 \text{ and } g(0) = f'(0)$$

If we apply the preceding theorem to this function we get

$$\sum_{n\geq 1}^{\mathcal{R}}\frac{f(n)}{n}-\sum_{n\geq 1}^{\mathcal{R}}\frac{f(0)}{n}=\int_0^1\frac{f(x)-f(0)}{x}dx-\frac{1}{2}f'(0)-\sum_{k=1}^{+\infty}\partial^k g(0)\frac{B_{k+1}}{(k+1)!}$$

Since $\partial^k g(0)=\frac{\partial^{k+1}f(0)}{k+1}$ we obtain the formula. $\square$

Examples

(1) For $z\in U_\pi$ (cf 1.15) take the entire function $x\mapsto e^{-zx}$. We have by the preceding corollary

$$\sum_{n\geq 1}^{\mathcal{R}}\frac{e^{-zn}}{n}=\int_0^1\frac{e^{-zx}-1}{x}dx+\gamma+\frac{1}{2}z-\sum_{k=2}^{+\infty}\frac{(-1)^k}{k!}z^k\frac{B_k}{k}$$

The sum $\sum_{k=2}^{+\infty}\frac{(-1)^k}{k!}z^k\frac{B_k}{k}$ is easily obtained by integration of the relation

$$\sum_{k=2}^{+\infty}\frac{(-1)^k}{k!}z^{k-1}B_k=\frac{e^{-z}}{1-e^{-z}}-\frac{1}{z}+\frac{1}{2}$$

which gives

$$\sum_{k=2}^{+\infty}\frac{(-1)^k}{k!}z^k\frac{B_k}{k}=Log(1-e^{-z})-Log(z)+\frac{1}{2}z$$

Finally we get for $z\in U_\pi$

$$\sum_{n\geq 1}^{\mathcal{R}}\frac{e^{-zn}}{n}=\gamma+\int_0^1\frac{e^{-zx}-1}{x}dx-Log(1-e^{-z})+Log(z)\qquad(2.12)$$

For $Re(z)>0$ or $z=i\alpha$ with $\alpha\in]-\pi,\pi[\setminus\{0\}$ we are in a case of convergence then

$$\sum_{n\geq 1}^{+\infty}\frac{e^{-zn}}{n}=\int_0^1\frac{e^{-zx}-1}{x}dx+\int_1^{+\infty}\frac{e^{-zx}}{x}dx+\gamma+Log(z)$$
$$-Log(1-e^{-z})$$

Since $\sum_{n\geq 1}^{+\infty} \frac{e^{-zn}}{n} = -Log(1 - e^{-z})$, this gives the relation

$$\int_1^{+\infty} \frac{e^{-zx}}{x}dx + \int_0^1 \frac{e^{-zx} - 1}{x}dx = -\gamma - Log(z) \tag{2.13}$$

Taking the real part of this relation for $z = i\alpha$ we obtain

$$\int_1^{+\infty} \frac{\cos(\alpha x)}{x}dx + \int_0^1 \frac{\cos(\alpha x) - 1}{x}dx = -\gamma - Log(|\alpha|) \tag{2.14}$$

(2) Let's take $Re(z) > 0$, if we apply the formula (2.4) of summation by parts to the functions $f(x) = Log(x)$ and $g(x) = e^{-zx}$, we get

$$\sum_{n\geq 1}^{\mathcal{R}} e^{-nz}Log(1 + \frac{1}{n}) = (e^z - 1)\sum_{n\geq 1}^{\mathcal{R}} e^{-nz}Log(n) + \int_0^1 Log(t + 1)e^{-zt}dt$$

2.3.2 An Expression of Catalan's Constant

Let f be an entire function of exponential type $< \pi$. By the same method of the preceding corollary we have

$$\sum_{n\geq 1}^{\mathcal{R}} \frac{f(n)}{n^2} = \int_0^1 \frac{f(x) - f(0) - f'(0)x}{x^2}dx + (\sum_{n\geq 1}^{\mathcal{R}} \frac{1}{n^2})f(0) + \gamma f'(0) - \frac{1}{4}f''(0)$$

$$- \sum_{k=2}^{+\infty} \frac{\partial^{k+1}f(0)}{(k+1)!} \frac{B_k}{k}$$

With $f(x) = \sin(\alpha x)$, $0 < \alpha < \pi$, we get

$$\sum_{n\geq 1}^{\mathcal{R}} \frac{\sin(\alpha n)}{n^2} = \int_0^1 \frac{\sin(\alpha x) - \alpha x}{x^2}dx + \gamma\alpha - \sum_{k=1}^{+\infty}(-1)^k \frac{\alpha^{2k+1}}{(2k+1)!} \frac{B_{2k}}{2k}$$

Since we are in a case of convergence we obtain

$$\sum_{n\geq 1}^{+\infty} \frac{\sin(\alpha n)}{n^2} = \int_0^1 \frac{\sin(\alpha x) - \alpha x}{x^2}dx + \int_1^{+\infty} \frac{\sin(\alpha x)}{x^2}dx + \gamma\alpha$$

$$- \sum_{k=1}^{+\infty}(-1)^k \frac{\alpha^{2k+1}}{(2k+1)!} \frac{B_{2k}}{2k}$$

By integration of (2.14) we get

$$\int_0^1 \frac{\sin(\alpha x) - \alpha x}{x^2} dx + \int_1^{+\infty} \frac{\sin(\alpha x)}{x^2} dx = \alpha - \alpha Log(\alpha) - \alpha\gamma \qquad (2.15)$$

Thus

$$\sum_{n \geq 1}^{+\infty} \frac{\sin(\alpha n)}{n^2} = \alpha - \alpha Log(\alpha) - \sum_{k=1}^{+\infty} (-1)^k \frac{\alpha^{2k+1}}{(2k+1)!} \frac{B_{2k}}{2k}$$

The Catalan's constant G is defined by

$$G = \sum_{n=1}^{\infty} \frac{(-1)^{n-1}}{(2n-1)^2} = \sum_{n=1}^{+\infty} \frac{\sin(\frac{\pi}{2}n)}{n^2}$$

thus we get the expression of the Catalan's constant:

$$G = \frac{\pi}{2} - \frac{\pi}{2} Log(\frac{\pi}{2}) - \sum_{k=1}^{+\infty} (-1)^k (\frac{\pi}{2})^{2k+1} \frac{B_{2k}}{2k(2k+1)!}$$

2.3.3 Some Trigonometric Series

Let f be an entire function of exponential type $< \pi$ and the partial sum of its Taylor expansion

$$T_p(f)(x) = f(0) + f'(0)x + \ldots + \frac{\partial^{p-1}f(0)}{(p-1)!} x^{p-1}$$

By the same method of the preceding corollary we have for any integer $p \geq 1$

$$\sum_{n \geq 1}^{\mathcal{R}} \frac{f(n)}{n^p} = \int_0^1 \frac{f(x) - T_p(f)(x)}{x^p} dx$$

$$+ (\sum_{n \geq 1}^{\mathcal{R}} \frac{1}{n^p}) f(0) + (\sum_{n \geq 1}^{\mathcal{R}} \frac{1}{n^{p-1}}) f'(0) + \ldots + \gamma \frac{\partial^{p-1}f(0)}{(p-1)!} - \frac{1}{2} \frac{\partial^p f(0)}{p!}$$

$$- \sum_{k=2}^{+\infty} \frac{\partial^{k+p-1}f(0)}{(k+p-1)!} \frac{B_k}{k}$$

thus in term of zeta values we obtain

$$\sum_{n\geq 1}^{\mathcal{R}} \frac{f(n)}{n^p} = \int_0^1 \frac{f(x) - T_p(f)(x)}{x^p} dx$$

$$+ \sum_{k=0}^{p-2} (\zeta(p-k) - \frac{1}{p-k-1}) \frac{\partial^k f(0)}{k!}$$

$$+ \gamma \frac{\partial^{p-1} f(0)}{(p-1)!} - \frac{1}{2} \frac{\partial^p f(0)}{p!}$$

$$- \sum_{k=2}^{+\infty} \frac{\partial^{k+p-1} f(0)}{(k+p-1)!} \frac{B_k}{k}$$

Example With $f(x) = \sin(\alpha x)$, $0 < \alpha < \pi$, and $p = 4$, we are in a case of convergence, thus we get

$$\sum_{n\geq 1}^{+\infty} \frac{\sin(\alpha n)}{n^4} = \int_0^1 \frac{\sin(\alpha x) - \alpha x + \alpha^3 \frac{x^3}{6}}{x^4} dx + \int_1^{+\infty} \frac{\sin(\alpha x)}{x^4} dx$$

$$+ \alpha(\zeta(3) - \frac{1}{2}) - \gamma \frac{\alpha^3}{6} - \sum_{k=1}^{+\infty} (-1)^{k-1} \frac{\alpha^{2k+3}}{(2k+3)!} \frac{B_{2k}}{2k}$$

By repeated integration of (2.15) we have

$$\int_0^1 \frac{\sin(\alpha x) - \alpha x + \alpha^3 \frac{x^3}{6}}{x^4} dx + \int_1^{+\infty} \frac{\sin(\alpha x)}{x^4} dx = \frac{1}{6}\alpha^3 (Log(\alpha) + \gamma - \frac{11}{6}) + \frac{1}{2}\alpha$$

Thus

$$\sum_{n\geq 1}^{+\infty} \frac{\sin(\alpha n)}{n^4} = \frac{1}{6}\alpha^3 (Log(\alpha) - \frac{11}{6}) + \alpha\zeta(3) - \sum_{k=1}^{+\infty} (-1)^{k-1} \frac{\alpha^{2k+3}}{(2k+3)!} \frac{B_{2k}}{2k}$$

Since for $\alpha = \frac{\pi}{2}$ we have

$$\sum_{n=1}^{\infty} \frac{(-1)^{n-1}}{(2n-1)^4} = \sum_{n=1}^{+\infty} \frac{\sin(\frac{\pi}{2}n)}{n^4}$$

we get

$$\sum_{n=1}^{\infty} \frac{(-1)^{n-1}}{(2n-1)^4} = \frac{1}{6}(\frac{\pi}{2})^3 (Log(\frac{\pi}{2}) - \frac{11}{6}) + \frac{\pi}{2}\zeta(3) - \sum_{k=1}^{+\infty} (-1)^{k-1} \frac{(\frac{\pi}{2})^{2k+3}}{(2k+3)!} \frac{B_{2k}}{2k}$$

By derivation with respect to α we get

$$\sum_{n\geq 1}^{+\infty} \frac{\cos(\alpha n)}{n^3} = \frac{1}{2}\alpha^2 Log(\alpha) - \frac{3}{4}\alpha^2 + \zeta(3) - \sum_{k=1}^{+\infty}(-1)^{k-1}\frac{\alpha^{2k+2}}{(2k+2)!}\frac{B_{2k}}{2k}$$

and taking $\alpha = \frac{\pi}{2}$ we have

$$\sum_{n\geq 1}^{+\infty} \frac{\cos(\frac{\pi}{2}n)}{n^3} = \sum_{n\geq 1}^{+\infty} \frac{(-1)^n}{(2n)^3} = -\frac{3}{32}\zeta(3)$$

thus we get

$$\zeta(3) = \frac{6}{35}\pi^2 - \frac{4}{35}\pi^2 Log(\frac{\pi}{2}) + \frac{32}{35}\sum_{k=1}^{+\infty}(-1)^{k-1}\frac{(\frac{\pi}{2})^{2k+2}}{(2k+2)!}\frac{B_{2k}}{2k}$$

(if we write B_{2k} in terms of $\zeta(2k)$ this is a special case of an identity given by Srivastava and Choi p.409 (22)).

Remark (Summation over $\mathbb{Z}$) Since Fourier series are often given by summations over $\mathbb{Z}$ it is natural to ask for the possibility of a Ramanujan summation over $\mathbb{Z}$ for an entire function $f \in \mathcal{O}^\pi$ such that the function $x \mapsto f(-x+1)$ is also in $\mathcal{O}^\pi$.

Then we can try to define $\sum_{n\in\mathbb{Z}}^{\mathcal{R}} f(n)$ by breaking the sum in two parts

$$\sum_{n\geq 1}^{\mathcal{R}} f(n) + \sum_{n\geq 1}^{\mathcal{R}} f(-n+1)$$

Now by the shift property we find easily that

$$\sum_{n\geq 1}^{\mathcal{R}} f(n+m) + \sum_{n\geq 1}^{\mathcal{R}} f(-n+1+m) - \int_m^{m+1} f(x)dx$$

is independent of m, thus *we can define $\sum_{n\in\mathbb{Z}}^{\mathcal{R}} f(n)$ by*

$$\sum_{n\in\mathbb{Z}}^{\mathcal{R}} f(n) = \sum_{n\geq 1}^{\mathcal{R}} f(n) + \sum_{n\geq 1}^{\mathcal{R}} f(-n+1) - \int_0^1 f(x)dx$$

$$= \sum_{n\geq 1}^{\mathcal{R}} f(n) + f(0) + \sum_{n\geq 1}^{\mathcal{R}} f(-n) - \int_{-1}^1 f(x)dx$$

For example, let's take $a \in \mathbb{C}$ and $|a| < \pi$ and $a \neq 0$, then the "divergent calculation" of Euler

$$\sum_{n\in\mathbb{Z}} e^{an} = \sum_{n\geq 1} e^{an} + e^a \sum_{n\geq 1} e^{-an} = \frac{e^a}{1-e^a} + \frac{1}{1-e^{-a}} = 0$$

is justified if we apply our preceding definition to the function $f : x \mapsto e^{ax}$ since

$$\sum_{n\in\mathbb{Z}}^{\mathcal{R}} e^{an} = \sum_{n\geq 1}^{\mathcal{R}} e^{an} + e^a \sum_{n\geq 1}^{\mathcal{R}} e^{-an} - \int_0^1 e^{ax}dx$$

$$= (\frac{e^a}{1-e^a} + \frac{e^a}{a}) + (\frac{1}{1-e^{-a}} - \frac{1}{a}) - \frac{e^a-1}{a}$$

$$= 0$$

Note also that for any integer $k \geq 0$ we have

$$\sum_{n\in\mathbb{Z}}^{\mathcal{R}} n^k = \sum_{n\geq 1}^{\mathcal{R}} n^k + \sum_{n\geq 1}^{\mathcal{R}} (-1)^k n^k - \int_{-1}^1 x^k dx = 0$$

Note that if we apply our preceding definition to the function $x \mapsto \frac{1}{z-x\pi}$ (which is not entire) we find that for $z \in \mathbb{C}\backslash\pi\mathbb{Z}$ we have

$$\sum_{n\in\mathbb{Z}}^{\mathcal{R}} \frac{1}{z-n\pi} = \sum_{n\geq 1}^{\mathcal{R}} \frac{1}{z-n\pi} + \sum_{n\geq 1}^{\mathcal{R}} \frac{1}{z+n\pi} + \frac{1}{z} - \int_{-1}^1 \frac{1}{z-x\pi}dx$$

$$= \sum_{n\geq 1}^{\mathcal{R}} \frac{2z}{z^2-n^2\pi^2} + \frac{1}{z} + \frac{1}{\pi}Log(\frac{z-\pi}{z+\pi})$$

$$= \sum_{n=1}^{+\infty} \frac{2z}{z^2-n^2\pi^2} - \int_1^{+\infty} \frac{2z}{z^2-x^2\pi^2}dx + \frac{1}{z} + \frac{1}{\pi}Log(\frac{z-\pi}{z+\pi})$$

$$= \cot(z)$$

2.4 Functional Relations for Fractional Sums

Let's take $f \in \mathcal{O}^\pi$ and an integer $N > 1$, we note $f(x/N)$ the function

$$x \mapsto f(\frac{x}{N})$$

Theorem 7 *Let's take $f \in \mathcal{O}^\pi$ and an integer $N > 1$, then*

$$R_{f(x/N)}(x) = \sum_{k=0}^{N-1} R_f(\frac{x+k}{N}) - N \int_{1/N}^1 f(x)dx \qquad (2.16)$$

And we get

$$\sum_{n\geq 1}^{\mathcal{R}} f(\frac{n}{N}) = \sum_{k=0}^{N-1} [\sum_{n\geq 1}^{\mathcal{R}} f(n - \frac{k}{N})] - \sum_{k=1}^{N-1} \int_{1}^{k/N} f(x)dx - N \int_{1/N}^{1} f(x)dx \qquad (2.17)$$

Proof The function

$$R(x) = \sum_{k=0}^{N-1} R_f(\frac{x+k}{N}) = R_f(\frac{x}{N}) + R_f(\frac{x+1}{N}) + \ldots + R_f(\frac{x+N-1}{N})$$

satisfies

$$R(x) - R(x+1) = R_f(\frac{x}{N}) - R_f(\frac{x}{N}+1) = f(\frac{x}{N})$$

therefore by definition of the fractional remainder we have

$$R_{f(\frac{x}{N})}(x) = \sum_{k=0}^{N-1} R_f(\frac{x+k}{N}) - \int_{1}^{2} \sum_{k=0}^{N-1} R_f(\frac{x+k}{N})dx$$

Since

$$\int_{1}^{2} \sum_{k=0}^{N-1} R_f(\frac{x+k}{N})dx = N \int_{\frac{1}{N}}^{\frac{1}{N}+1} R_f(x)dx = N \int_{1/N}^{1} f(x)dx$$

we get

$$R_{f(x/N)}(x) = \sum_{k=0}^{N-1} R_f(\frac{x+k}{N}) - N \int_{1/N}^{1} f(x)dx$$

and for $x = 1$ we deduce that

$$\sum_{n\geq 1}^{\mathcal{R}} f(\frac{n}{N}) = \sum_{k=0}^{N-1} R_f(\frac{k+1}{N}) - N \int_{1/N}^{1} f(x)dx$$

We have by (2.2)

$$R_f(\frac{k+1}{N}) = \sum_{n\geq 1}^{\mathcal{R}} f(n - 1 + \frac{k+1}{N}) - \int_{1}^{(k+1)/N} f(x)dx$$

thus we get

$$\sum_{n\geq 1}^{\mathcal{R}} f(\frac{n}{N}) = \sum_{k=0}^{N-1} \sum_{n\geq 1}^{\mathcal{R}} f(n-1+\frac{k+1}{N}) - \sum_{k=1}^{N-1} \int_1^{k/N} f(x)dx - N\int_{1/N}^1 f(x)dx$$

which is

$$\sum_{n\geq 1}^{\mathcal{R}} f(\frac{n}{N}) = \sum_{k=0}^{N-1} \sum_{n\geq 1}^{\mathcal{R}} f(n-\frac{k}{N}) - \sum_{k=1}^{N-1} \int_1^{k/N} f(x)dx - N\int_{1/N}^1 f(x)dx$$

□

Remark If $g \in \mathcal{O}^{\pi/N}$, then with $f : x \mapsto g(Nx)$ we have

$$\sum_{n\geq 1}^{\mathcal{R}} g(n) = \sum_{k=0}^{N-1} [\sum_{n\geq 1}^{\mathcal{R}} g(Nn-k)] + \sum_{k=1}^{N-1} \frac{1}{N} \int_k^N g(x)dx - \int_1^N g(x)dx$$

Thus if $N = 2$, we get for $f \in \mathcal{O}^{\pi/2}$

$$\sum_{n\geq 1}^{\mathcal{R}} f(n) = \sum_{n\geq 1}^{\mathcal{R}} f(2n) + \sum_{n\geq 1}^{\mathcal{R}} f(2n-1) - \frac{1}{2}\int_1^2 f(x)dx \qquad (2.18)$$

Example For $f(x) = \frac{1}{x}$ we get by (2.18)

$$\sum_{n\geq 1}^{\mathcal{R}} \frac{1}{n} = \frac{1}{2} \sum_{n\geq 1}^{\mathcal{R}} \frac{1}{n} + \sum_{n\geq 1}^{\mathcal{R}} \frac{1}{2n-1} - \frac{1}{2}Log(2)$$

thus

$$\sum_{n\geq 1}^{\mathcal{R}} \frac{1}{2n-1} = \frac{\gamma}{2} + \frac{1}{2}Log(2)$$

and by the shift property

$$\sum_{n\geq 1}^{\mathcal{R}} \frac{1}{2n+1} = -1 + \frac{\gamma}{2} + \frac{1}{2}Log(2) + \frac{1}{2}Log(3)$$

This leads to an easy solution to a question submitted by Ramanujan in the Journal of Indian Mathematical Society, that is how to prove that

$$1 + 2 \sum_{n=1}^{+\infty} \frac{1}{(4n)^3 - 4n} = \frac{3}{2} Log(2)$$

This relation is a simple consequence of the fact that

$$\sum_{n=1}^{\mathcal{R}} \frac{1}{(4n)^3 - 4n} = -\frac{1}{4} \sum_{n=1}^{\mathcal{R}} \frac{1}{n} + \frac{1}{2} \left(\sum_{n=1}^{\mathcal{R}} \frac{1}{4n - 1} + \sum_{n=1}^{\mathcal{R}} \frac{1}{4n + 1} \right)$$

With $f(x) = \frac{1}{2x+1}$, formula (2.18) gives

$$\sum_{n \geq 1}^{\mathcal{R}} \frac{1}{4n + 1} + \sum_{n \geq 1}^{\mathcal{R}} \frac{1}{4n - 1} = \sum_{n \geq 1}^{\mathcal{R}} \frac{1}{2n + 1} + \frac{1}{4}(Log(5) - Log(3))$$

$$= -1 + \frac{\gamma}{2} + \frac{1}{2} Log(2) + \frac{1}{4} Log(3) + \frac{1}{4} Log(5)$$

thus

$$\sum_{n=1}^{\mathcal{R}} \frac{1}{(4n)^3 - 4n} = -\frac{1}{2} + \frac{1}{4} Log(2) + \frac{1}{8} Log(3) + \frac{1}{8} Log(5)$$

Since we are in a case of convergence then

$$\sum_{n=1}^{+\infty} \frac{1}{(4n)^3 - 4n} = \sum_{n=1}^{\mathcal{R}} \frac{1}{(4n)^3 - 4n} + \int_1^{+\infty} \frac{1}{(4x)^3 - 4x} dx = -\frac{1}{2} + \frac{3}{4} Log(2)$$

Theorem 8 *Let's take $f \in \mathcal{O}^\pi$ and an integer $N > 1$, then*

$$\varphi_{f(x/N)}(x) = \sum_{j=0}^{N-1} \varphi_f\left(\frac{x - j}{N}\right) + \sum_{n \geq 1}^{\mathcal{R}} f\left(\frac{n}{N}\right) - N \sum_{n \geq 1}^{\mathcal{R}} f(n) + N \int_{1/N}^1 f(x)dx$$

which is the entry 7 Ch VI of Ramanujan's Notebook (corrected with the addition of an integral term).

Proof We can write (2.5) in the form

$$R_{f(x/N)}(x + 1) = R_f\left(\frac{x + 1}{N}\right) + R_f\left(\frac{x + 2}{N}\right) + \ldots + R_f\left(\frac{x + N}{N}\right) - N \int_{1/N}^1 f(x)dx$$

with $R_{f(x/N)}(x+1) = \sum_{n\geq 1}^{\mathcal{R}} f(\frac{n}{N}) - \varphi_{f(x/N)}(x)$ we get

$$\sum_{n\geq 1}^{\mathcal{R}} f(\frac{n}{N}) - \varphi_{f(x/N)}(x) = N \sum_{n\geq 1}^{\mathcal{R}} f(n)$$

$$- [\varphi_f(\frac{x+1-N}{N}) + \ldots + \varphi_f(\frac{x+N-N}{N})]$$

$$- N \int_{1/N}^1 f(x)dx$$

$\square$

Corollary *Since $\varphi_{f(x/N)}(0) = 0$ and $\varphi_f(0) = 0$ then we get*

$$\sum_{k=1}^{N-1} \varphi_f(\frac{-k}{N}) = N \sum_{n\geq 1}^{\mathcal{R}} f(n) - \sum_{n\geq 1}^{\mathcal{R}} f(\frac{n}{N}) - N \int_{1/N}^1 f(x)dx$$

a formula that Ramanujan gives without the correcting integral term.

Examples

(1) If $f(x) = \frac{1}{x}$, we have $f(x/N) = Nf(x)$ and since $\varphi_f(x) = \gamma + \psi(x+1)$, the preceding theorem gives

$$N(\gamma + \psi(x+1)) = N\gamma + N(\psi(\frac{x+1}{N}) + \ldots + \Psi(\frac{x+N}{N})) + NLog(N)$$

and we get the well known formula

$$\Psi(x) = \frac{1}{N} \sum_{k=0}^{N-1} \psi(\frac{x+k}{N}) + Log(N)$$

(2) If $f(x) = Log(x)$, then $\varphi_f(x) = Log(\Gamma(x+1))$ and

$$\varphi_{f(x/N)}(x) = \varphi_{Log}(x) - \varphi_{Log(N)}(x) = Log(\Gamma(x+1)) - xLog(N)$$

With the preceding theorem we get

$$Log(\Gamma(x+1)) = \sum_{k=1}^{N} Log(\frac{\Gamma(\frac{x+k}{N})}{\sqrt{2\pi}}) + (x + \frac{1}{2})Log(N) + Log(\sqrt{2\pi})$$

Taking the exponential we get the Gauss formula for the Gamma function.

$$\Gamma(Nx) = (2\pi)^{\frac{1-N}{2}} N^{Nx-\frac{1}{2}} \prod_{k=0}^{N-1} \Gamma(x + \frac{k}{N})$$

(3) If $f(x) = \frac{Log(x)}{x}$, then $f(x/N) = N\frac{Log(x)}{x} - NLog(N)\frac{1}{x}$, thus

$$\varphi_{f(x/N)}(x) = N\varphi_f(x) - NLog(N)(\gamma + \psi(x+1))$$

and we get

$$N\varphi_f(x) - NLog(N)\psi(x+1) = \sum_{j=0}^{N-1} \varphi_f(\frac{x-j}{N}) + N\int_{1/N}^{1} f(x)dx$$

this gives entry 17 iii of Chap. 8 in Ramanujan's Notebook:

$$\varphi_{\frac{Log(x)}{x}}(x) = \frac{1}{N}\sum_{j=0}^{N-1} \varphi_{\frac{Log(x)}{x}}(\frac{x-j}{N}) + Log(N)\psi(x+1) - \frac{1}{2}Log^2(N)$$

(4) If $f(x) = Log^2(x)$, then $f(x/N) = Log^2(x) - 2Log(x)Log(N) + Log^2(N)$, thus

$$\varphi_{f(x/N)}(x) = \varphi_f(x) - 2Log(N)Log(\Gamma(x+1)) + xLog^2(N)$$

and by the preceding theorem

$$\varphi_{f(x/N)}(x) = \sum_{j=0}^{N-1} \varphi_f(\frac{x-j}{N})$$

$$+ (1-N)\sum_{n\geq 1}^{\mathcal{R}} Log^2(n) - 2Log(N)(Log(\sqrt{2\pi}) - 1) + \frac{1}{2}Log^2(N)$$

$$+ N\int_{1/N}^{1} f(x)dx$$

this gives

$$\varphi_f(x) = \sum_{j=0}^{N-1} \varphi_f(\frac{x-j}{N}) + 2Log(N)Log(\frac{\Gamma(x+1)}{\sqrt{2\pi}}) - (\frac{1}{2} + x)Log^2(N)$$

$$- (N-1)(\sum_{n\geq 1}^{\mathcal{R}} Log^2(n) - 2)$$

this is 18(ii) of Chap. 8 in Ramanujan's Notebook (with $C = \sum_{n\geq 1}^{\mathcal{R}} Log^2(n) - 2$).

Chapter 3
Dependence on a Parameter

In this chapter we give three fundamental results on the Ramanujan summation of series depending on a parameter.

In the first section we prove that the Ramanujan summation conserves the property of analyticity with respect to an external parameter z. We deduce the result that an expansion of $f(x, z)$ in terms of a power series in z gives the corresponding expansion of the sum $\sum_{n\geq1}^{\mathcal{R}} f(n, z)$. Some consequences of this result are examined.

In the second section we study the interchange of the Ramanujan summation and integration with respect to a parameter $u \in I$ where $I \subset \mathbb{R}$ is a given interval. This gives a simple integral formula for the Ramanujan summation of a Laplace transform.

In the third section we prove that with a very simple hypothesis we can interchange the two Ramanujan summations in $\sum_{n\geq1}^{\mathcal{R}} \sum_{m\geq1}^{\mathcal{R}} f(m, n)$. As a consequence we easily prove a functional relation for Eisenstein function G_2.

3.1 Analyticity with Respect to a Parameter

3.1.1 The Theorem of Analyticity

It is well known that the simple convergence of a series $\sum_{n\geq1} f(n, z)$ of analytic functions for z in a domain U does not imply that the sum $\sum_{n\geq1}^{+\infty} f(n, z)$ is analytic in U.

A very important property of the Ramanujan summation is that analyticity of the terms implies analyticity of the sum.

© Springer International Publishing AG 2017   61
B. Candelpergher, *Ramanujan Summation of Divergent Series*,
Lecture Notes in Mathematics 2185, DOI 10.1007/978-3-319-63630-6_3

We have an illustration of this fact with

$$\sum_{n\geq 1}^{\mathcal{R}} \frac{1}{n^z} = \zeta(z) - \frac{1}{z-1}$$

where we see that the pole of ζ is removed.

Definition 5 Let $(x, z) \mapsto f(x, z)$ be a function defined for $Re(x) > 0$ and $z \in U \subset \mathbb{C}$ such that $z \mapsto f(x, z)$ is analytic for $z \in U$. We say that f is *locally uniformly in* $\mathcal{O}^\pi$ if

(a) for all $z \in U$ the function $x \mapsto f(x, z)$ is analytic for $Re(x) > 0$
(b) for any K compact of U there exist $\alpha < \pi$ and $C > 0$ such that for $Re(x) > 0$ and $z \in K$

$$|f(x, z)| \leq Ce^{\alpha|x|}$$

By the Cauchy formula there is the same type of inequality for the derivatives $\partial_z^k f(x, z)$ thus for any integer $k \geq 1$ the function $(x, z) \mapsto \partial_z^k f(x, z)$ is also locally uniformly in $\mathcal{O}^\pi$. Thus we note that for any integer $k \geq 1$ the sum $\sum_{n\geq 1}^{\mathcal{R}} \partial_z^k f(n, z)$ is well defined.

Theorem 9 (Analyticity of $z \mapsto \sum_{n\geq 1}^{\mathcal{R}} f(n, z)$) *Let $(x, z) \mapsto f(x, z)$ be a function defined for $Re(x) > 0$ and $z \in U \subset \mathbb{C}$ such that $z \mapsto f(x, z)$ is analytic for $z \in U$ and f is locally uniformly in $\mathcal{O}^\pi$. Then the function*

$$z \mapsto \sum_{n\geq 1}^{\mathcal{R}} f(n, z)$$

is analytic in U and

$$\partial_z^k \sum_{n\geq 1}^{\mathcal{R}} f(n, z) = \sum_{n\geq 1}^{\mathcal{R}} \partial_z^k f(n, z)$$

Thus if $z_0 \in U$ and

$$f(n, z) = \sum_{k=0}^{+\infty} a_k(n)(z - z_0)^k \text{ for } |z - z_0| < \rho \text{ and } n \geq 1$$

then

$$\sum_{n\geq 1}^{\mathcal{R}} f(n, z) = \sum_{k=0}^{+\infty} \left[\sum_{n\geq 1}^{\mathcal{R}} a_k(n)\right](z - z_0)^k$$

Proof Let z be in a compact $K \subset U$, then by definition we have

$$\sum_{n \geq 1}^{\mathcal{R}} f(n, z) = \frac{f(1, z)}{2} + iJ_f(z)$$

with

$$J_f(z) = \int_0^{+\infty} \frac{f(1 + it, z) - f(1 - it, z)}{e^{2\pi t} - 1} dt$$

For every $t \in]0, +\infty[$ the function

$$z \mapsto \frac{f(1 + it, z) - f(1 - it, z)}{e^{2\pi t} - 1}$$

is analytic in U and, if $z \in K$, we have by hypothesis a constant C such that

$$\left| \frac{f(1 + it, z) - f(1 - it, z)}{e^{2\pi t} - 1} \right| \leq 2Ce^\alpha \frac{e^{\alpha t}}{e^{2\pi t} - 1}$$

then by the analyticity theorem of an integral depending on a parameter, we get the analyticity of J_f in U and for $t > 1$ we have

$$\partial J_f(z) = \int_0^{+\infty} \frac{\partial_z f(1 + it, z) - \partial_z f(1 - it, z)}{e^{2\pi t} - 1} dt = J_{\partial_z f}$$

Thus the function $z \mapsto \sum_{n \geq 1}^{\mathcal{R}} f(n, z)$ is analytic in U and

$$\partial_z \sum_{n \geq 1}^{\mathcal{R}} f(n, z) = \partial_z \frac{f(1, z)}{2} + i \, \partial_z J_f(z) = \partial_z \frac{f(1, z)}{2} + iJ_{\partial_z f}(z)$$

that is

$$\partial_z \sum_{n \geq 1}^{\mathcal{R}} f(n, z) = \sum_{n \geq 1}^{\mathcal{R}} \partial_z f(n, z)$$

Repeated application of this procedure gives for any integer $k \geq 1$

$$\partial_z^k \sum_{n \geq 1}^{\mathcal{R}} f(n, z) = \sum_{n \geq 1}^{\mathcal{R}} \partial_z^k f(n, z)$$

For any $z_0 \in U$ we have the power series expansion

$$f(n, z) = \sum_{k=0}^{+\infty} a_k(n)(z - z_0)^k \text{ for } |z - z_0| < \rho \text{ and } n \geq 1$$

where $a_k(n) = \frac{1}{k!} \partial_z^k f(n, z_0)$.

Since the function $g : z \mapsto \sum_{n \geq 1}^{\mathcal{R}} f(n, z)$ is analytic in U we also have

$$g(z) = \sum_{k=0}^{+\infty} \frac{1}{k!} \partial^k g(z_0)(z - z_0)^k$$

and by the preceding result

$$\partial^k g(z_0) = \sum_{n \geq 1}^{\mathcal{R}} \partial_z^k f(n, z_0)$$

Thus we have

$$\sum_{n \geq 1}^{\mathcal{R}} f(n, z) = \sum_{k=0}^{+\infty} \sum_{n \geq 1}^{\mathcal{R}} \frac{1}{k!} \partial_z^k f(n, z_0)(z - z_0)^k$$

$$= \sum_{k=0}^{+\infty} [\sum_{n \geq 1}^{\mathcal{R}} a_k(n)](z - z_0)^k$$

$\square$

Corollary 10 *Consider a function f analytic for $\mathrm{Re}(x) > 0$ and of moderate growth, then for $z > 0$ we have*

$$\lim_{z \to 0+} \left(\sum_{n \geq 1}^{+\infty} f(n) e^{-nz} - \int_1^{+\infty} f(x) e^{-xz} \right) = \sum_{n \geq 1}^{\mathcal{R}} f(n)$$

Proof By the preceding theorem if $(x, z) \mapsto g(x, z)$ is a function uniformly in $\mathcal{O}^{\pi}$ for z in a neighborhood U_{z_0} of $z_0 \in \mathbb{C}$, then

$$\lim_{z \to z_0} \sum_{n \geq 1}^{\mathcal{R}} g(n, z) = \sum_{n \geq 1}^{\mathcal{R}} \lim_{z \to z_0} g(n, z)$$

As a special case we consider $g(x, z) = f(x)e^{-xz}$ where f is of moderate growth, then with $z_0 = 0$ we get

$$\lim_{z \to 0} \sum_{n \geq 1}^{\mathcal{R}} f(n)e^{-nz} = \sum_{n \geq 1}^{\mathcal{R}} f(n)$$

For $z > 0$ we are in a case of convergence, thus we get

$$\lim_{z \to 0+} \left(\sum_{n \geq 1}^{+\infty} f(n)e^{-nz} - \int_{1}^{+\infty} f(x)e^{-xz} \right) = \sum_{n \geq 1}^{\mathcal{R}} f(n)$$

$\square$

Examples

(1) Take $f(x, z) = \frac{1}{z+x}$ and $U = \{|z| < 1\}$, we have for $z \in U$ and $n \geq 1$

$$f(n, z) = \frac{1}{n} \frac{1}{1 + \frac{z}{n}} = \frac{1}{n} + \sum_{k=1}^{+\infty} \frac{(-1)^k}{n^{k+1}} z^k$$

thus by the preceding theorem with $z_0 = 0$ we have

$$\sum_{n \geq 1}^{\mathcal{R}} \frac{1}{z+n} = \gamma + \sum_{k=1}^{+\infty} (-1)^k (\zeta(k+1) - \frac{1}{k}) z^k$$

By (1.34) we get for $|z| < 1$

$$-\psi(z+1) + Log(z+1) = \gamma + \sum_{k=1}^{+\infty} (-1)^k \zeta(k+1) z^k - \sum_{k=1}^{+\infty} (-1)^k \frac{1}{k} z^k$$

thus we have the classical result

$$\psi(z+1) = -\gamma + \sum_{k=1}^{+\infty} (-1)^{k-1} \zeta(k+1) z^k$$

Note that by integrating this relation we get

$$Log(\Gamma(z+1)) = -\gamma z + \sum_{k=2}^{+\infty} (-1)^k \zeta(k) \frac{z^k}{k}$$

and since the series $\sum_{k\geq 2}(-1)^k\frac{\zeta(k)}{k}$ is convergent then, by the classical Abel theorem on power series applied with $z \to 1$, we get

$$\gamma = \sum_{k=2}^{+\infty}(-1)^{k-1}\frac{\zeta(k)}{k}$$

(2) Take $f : (x, s) \mapsto \frac{1}{x^s}$, the function $s \mapsto f(x, s)$ is analytic for $s \in \mathbb{C}$ and f is locally uniformly in $\mathcal{O}^\pi$. By the preceding theorem the function $s \mapsto \sum_{n\geq 1}^{\mathcal{R}}\frac{1}{n^s}$ is an entire function.

We have seen in Chap. 1 that $\sum_{n\geq 1}^{\mathcal{R}}\frac{1}{n^s} = \zeta(s) - \frac{1}{s-1}$ for $Re(s) > 1$, thus by analytic continuation we have

$$\sum_{n\geq 1}^{\mathcal{R}}\frac{1}{n^s} = \zeta(s) - \frac{1}{s-1} \quad \text{for } s \neq 1$$

If $s = -k$ with k integer ≥ 1 we get

$$\zeta(-k) + \frac{1}{k+1} = \sum_{n\geq 1}^{\mathcal{R}}n^k = \frac{1 - B_{k+1}}{k+1} \text{ if } k \geq 1$$

thus

$$\zeta(-k) = -\frac{B_{k+1}}{k+1} \text{ if } k \geq 1$$

and for $k = 0$

$$\zeta(0) = -1 + \sum_{n\geq 1}^{\mathcal{R}}1 = -\frac{1}{2}$$

By derivation with respect to the parameter s we get

$$\sum_{n\geq 1}^{\mathcal{R}}\frac{Log^k(n)}{n^s} = (-1)^k\partial^k\zeta(s) - \frac{k!}{(s-1)^{k+1}} \text{ for } s \neq 1.$$

In the case $k = 1, s = 0$, this gives

$$\zeta'(0) = -\sum_{n\geq 1}^{\mathcal{R}}Log(n) - 1 = -Log(\sqrt{2\pi})$$

For $s = 1$ we have the sums $\sum_{n\geq 1}^{\mathcal{R}} \frac{Log^k(n)}{n}$ which are related to the *Stieltjes constants* γ_k defined by the Laurent expansion of ζ at 1

$$\zeta(s+1) = \frac{1}{s} + \sum_{k\geq 0} \frac{(-1)^k}{k!} \gamma_k s^k$$

The expansion

$$\frac{1}{n^{s+1}} = \sum_{k\geq 0} \frac{(-1)^k}{k!} s^k \frac{Log^k(n)}{n}$$

gives by the preceding theorem

$$\sum_{n\geq 1}^{\mathcal{R}} \frac{1}{n^{s+1}} = \sum_{k\geq 0} \frac{(-1)^k}{k!} s^k \sum_{n\geq 1}^{\mathcal{R}} \frac{Log^k(n)}{n}$$

thus the Stieltjes constants are given by

$$\gamma_k = \sum_{n\geq 1}^{\mathcal{R}} \frac{Log^k(n)}{n}$$

Note that

$$\sum_{n\geq 1}^{\mathcal{R}} Log^2(n) = \zeta''(0) + 2$$

since $\zeta''(0)$ can be simply obtained by derivation of the functional equation of ζ (Berndt 1985, p.204) we see that the Stieltjes constant γ_1 is related to this sum by

$$\sum_{n\geq 1}^{\mathcal{R}} Log^2(n) = \gamma_1 + \frac{1}{2}\gamma^2 - \frac{1}{2}Log^2(2\pi) - \frac{\pi^2}{24} + 2 \tag{3.1}$$

(3) In the first page of chapter XV of his second Notebook Ramanujan writes (with a little change of notation):

$$\frac{\gamma + Log(z)}{z} + e^{-z}Log(1) + e^{-2z}Log(2) + \ldots = \frac{1}{2}Log(2\pi) \text{ when } z \text{ vanishes.}$$

The function

$$z \mapsto \sum_{n=1}^{+\infty} e^{-zn} Log(n)$$

is well defined when $Re(z) > 0$ and Berndt (Berndt 1985) gives an asymptotic expansion of this function when $z \to 0+$ by use of the Mellin inversion. Now we can give an exact formula for this sum by use of the Ramanujan summation.

Let $f(x, z) = e^{-zx} Log(x)$, for $z \in U_\pi$ (cf. Example 1 of Sect. 1.4.1). the function $x \mapsto f(x, z)$ is locally uniformly in $\mathcal{O}^\pi$ and

$$f(n, z) = \sum_{j=0}^{+\infty} \frac{(-1)^j}{j!} z^j n^j Log(n)$$

By the preceding example we have

$$\sum_{n \geq 1}^{\mathcal{R}} n^j Log(n) = -\zeta'(-j) - \frac{1}{(j+1)^2}$$

Thus by the theorem of analyticity we get for $z \in U_\pi$

$$\sum_{n \geq 1}^{\mathcal{R}} e^{-zn} Log(n) = \sum_{k=0}^{+\infty} \frac{(-1)^{k-1}}{k!} z^k (\zeta'(-k) + \frac{1}{(k+1)^2})$$

$$= \sum_{k=0}^{+\infty} \frac{(-1)^{k-1}}{k!} z^k (\zeta'(-k) + \sum_{k=1}^{+\infty} \frac{(-1)^k}{k!} \frac{z^{k-1}}{k}$$

$$= \sum_{k=0}^{+\infty} \frac{(-1)^{k-1}}{k!} z^k \zeta'(-k) + \frac{1}{z} \int_0^1 \frac{e^{-zx} - 1}{x} dx$$

For $Re(z) > 0$ and $-\pi < Im(z) < \pi$, we are in a case of convergence thus we have

$$\sum_{n \geq 1}^{\mathcal{R}} e^{-zn} Log(n) = \sum_{n=1}^{+\infty} e^{-zn} Log(n) - \int_1^{+\infty} e^{-zx} Log(x) dx$$

and using integration by parts

$$\sum_{n=1}^{+\infty} e^{-zn} Log(n) = \sum_{n \geq 1}^{\mathcal{R}} e^{-zn} Log(n) + \frac{1}{z} \int_1^{+\infty} \frac{e^{-zx}}{x} dx$$

Thus by formula (2.13) we get for $Re(z) > 0$ and $-\pi < Im(z) < \pi$

$$\sum_{n=1}^{+\infty} e^{-zn} Log(n) = \sum_{k=0}^{+\infty} \frac{(-1)^{k-1}}{k!} z^k \zeta'(-k) - \frac{\gamma + Log(z)}{z}$$

Since $\zeta'(0) = -Log(\sqrt{2\pi})$ we get the precise form of Ramanujan's formula

$$\frac{\gamma + Log(z)}{z} + \sum_{n=1}^{+\infty} e^{-zn} Log(n) = Log(\sqrt{2\pi}) + \sum_{k=1}^{+\infty} \frac{(-1)^{k-1}}{k!} z^k \zeta'(-k)$$

(4) Take the series $\sum_{n\geq 1} e^{-zn^\alpha}$ with $0 < \alpha \leq 1$ and $z \in \mathbb{C}$ with $Re(z) > 0$ and $-\pi < Im(z) < \pi$.

Using the power series expansion $e^{-zn^\alpha} = \sum_{k=0}^{+\infty} \frac{(-1)^k z^k}{k!} n^{\alpha k}$ the preceding theorem gives

$$\overset{\mathcal{R}}{\underset{n\geq 1}{\sum}} e^{-zn^\alpha} = \sum_{k=0}^{+\infty} \frac{(-1)^k z^k}{k!} \left(\zeta(-\alpha k) + \frac{1}{\alpha k + 1} \right)$$

Since we are in a case of convergence, we deduce that

$$\sum_{n=1}^{+\infty} e^{-zn^\alpha} = \sum_{k=0}^{+\infty} \frac{(-1)^k z^k}{k!} \zeta(-\alpha k) + \sum_{k=0}^{+\infty} \frac{(-1)^k z^k}{k!} \frac{1}{\alpha k + 1} + \int_1^{+\infty} e^{-zx^\alpha} dx$$

Now we observe that

$$\sum_{k=0}^{+\infty} \frac{(-1)^k z^k}{k!} \frac{1}{\alpha k + 1} + \int_1^{+\infty} e^{-zx^\alpha} dx = \int_0^{+\infty} e^{-zx^\alpha} dx = \frac{1}{\alpha} \Gamma(\frac{1}{\alpha}) z^{-\frac{1}{\alpha}}$$

thus we get for $0 < \alpha \leq 1$

$$\sum_{n=1}^{+\infty} e^{-zn^\alpha} = \sum_{k=0}^{+\infty} \frac{(-1)^k z^k}{k!} \zeta(-\alpha k) + \frac{1}{\alpha} \Gamma(\frac{1}{\alpha}) z^{-\frac{1}{\alpha}}$$

Note that this formula is not valid for $\alpha > 1$ since for $\alpha = 2$ it gives the formula

$$\sum_{n=1}^{+\infty} e^{-zn^2} = \zeta(0) + \frac{1}{2} \Gamma(\frac{1}{2}) z^{-\frac{1}{2}}$$

which is wrong since it is well known that the true formula involves exponentially small terms when $z \to 0$ (Bellman 1961):

$$\sum_{n=1}^{+\infty} e^{-zn^2} = \zeta(0) + \frac{1}{2} \Gamma(\frac{1}{2}) z^{-\frac{1}{2}} + \sqrt{\pi} z^{-\frac{1}{2}} \sum_{n=1}^{+\infty} e^{-\pi^2 n^2/z}$$

Remark With the same hypothesis as in the preceding theorem we have

$$R_{f(x,z)}(x) = \sum_{n\geq 1}^{\mathcal{R}} f(n+x,z) + f(x,z) - \int_1^{x+1} f(t,z)dt$$

$$\varphi_{f(x,z)}(x) = \sum_{n\geq 1}^{\mathcal{R}} f(n,z) - \sum_{n\geq 1}^{\mathcal{R}} f(n+x,z) + \int_1^{x+1} f(u,z)du$$

By the preceding theorem we get the analyticity of these functions of z and by derivation with respect to z we get

$$\partial_z R_{f(x,z)}(x) = R_{\partial_z f(x,z)}(x)$$

$$\partial_z \varphi_{f(x,z)}(x) = \varphi_{\partial_z f(x,z)}(x)$$

For example let $f(x,z) = \frac{1}{x^z}$ with $z \neq 1$ then

$$\varphi_{\frac{Log(x)}{x^z}} = -\partial_z \varphi_{\frac{1}{x^z}} = -\zeta'(z) + \partial_z \zeta(z,x) + \frac{Log(x)}{x^z}$$

For $z = 0$ then

$$\varphi_{Log}(x) = -\zeta'(0) + \partial_z \zeta(0,x) + Log(x)$$

but we know that $\varphi_{Log}(x) = Log(\Gamma(x+1))$ thus we get the Lerch formula (Berndt 1985)

$$Log(\Gamma(x)) = -\zeta'(0) + \partial_z \zeta(0,x)$$

3.1.2   Analytic Continuation of Dirichlet Series

Let a function $x \mapsto c(x)$, be analytic for $Re(x) > 0$, with an asymptotic expansion at infinity

$$c(x) = \sum_{k\geq 0} \alpha_k \frac{1}{x^{j_k}}$$

where $Re(j_0) < Re(j_1) < Re(j_2) < \ldots < Re(j_k) < \ldots$
   The Dirichlet series

$$h(s) = \sum_{n=1}^{+\infty} \frac{c(n)}{n^s}$$

defines for $Re(s) > 1 - Re(j_0)$ an analytic function h and since we are in a case of convergence we have

$$h(s) = \sum_{n \geq 1}^{\mathcal{R}} \frac{c(n)}{n^s} + \int_1^{+\infty} c(x) x^{-s} dx$$

By the theorem of analyticity the function $s \mapsto \sum_{n \geq 1}^{\mathcal{R}} \frac{c(n)}{n^s}$ is an entire function, thus the singularities of the function h are given by the integral term. Let us write formally

$$\int_1^{+\infty} c(x) x^{-s} dx = \int_1^{+\infty} \sum_{k \geq 0} \alpha_k x^{-s-j_k} dx$$

$$= \sum_{k \geq 0} \int_1^{+\infty} \alpha_k x^{-s-j_k} dx$$

$$= \sum_{k \geq 0} \alpha_k \frac{1}{s + j_k - 1}$$

In some cases this can give a simple proof that the function h has simple poles at the points $s = 1 - j_k$ with residues α_k.

Examples

(1) Take $h(s) = \sum_{n=1}^{+\infty} \frac{1}{(n+1)n^s}$ for $Re(s) > 0$, solution of

$$h(s) + h(s - 1) = \zeta(s)$$

We have for $Re(s) > 0$

$$h(s) = \sum_{n \geq 1}^{\mathcal{R}} \frac{1}{(n + 1)n^s} + \int_1^{+\infty} \frac{1}{(x + 1)x^s} dx$$

From the analyticity theorem we deduce that the function $s \mapsto \sum_{n \geq 1}^{\mathcal{R}} \frac{1}{(n+1)n^s}$ is an entire function, thus the singularities of the (analytic continuation of the) function h are the singularities of the function

$$k : s \mapsto \int_1^{+\infty} \frac{1}{(x + 1)x^s} dx$$

The analytic continuation of the function k is obtained simply by observing that for $x > 1$ we have

$$\frac{1}{(x + 1)x^s} = \sum_{k=0}^{+\infty} \frac{(-1)^k}{x^{s+k+1}}$$

and the dominated convergence theorem gives

$$k(s) = \int_1^{+\infty} \frac{1}{(x+1)x^s}dx = \sum_{k=0}^{+\infty}(-1)^k \int_1^{+\infty} x^{-s-k-1}dx = \sum_{k=0}^{+\infty}\frac{(-1)^k}{s+k}$$

Thus the function h has simple poles at $s = -k$, $k = 0, 1, 2, \ldots$, with residues $(-1)^k$ and

$$\lim_{s=-k}(h(s) - \frac{(-1)^k}{s+k}) = \sum_{n\geq 1}^{\mathcal{R}}\frac{n^k}{n+1} + \sum_{j\neq k}^{+\infty}\frac{(-1)^j}{j-k}$$

(2) Take $c(x) = \psi(x+1) + \gamma$ then $c(n) = H_n$ and for $\text{Re}(s) > 1$

$$h(s) = \sum_{n=1}^{+\infty}\frac{H_n}{n^s}$$

We have for $\text{Re}(s) > 1$

$$h(s) = \sum_{n\geq 1}^{\mathcal{R}}\frac{H_n}{n^s} + \int_1^{+\infty}(\psi(x+1) + \gamma)x^{-s}dx$$

The singularities of the integral term are easily obtained by the asymptotic expansion

$$\psi(x+1) + \gamma = Log(x) + \gamma + \frac{1}{2x} - \sum_{k\geq 1}\frac{B_{2k}}{2k}\frac{1}{x^{2k}}$$

This gives for h a pole of order 2 for $s = 1$ with residue γ, a simple pole for $s = 0$ with residue $1/2$ and simple poles for $s = 1 - 2k$ with residues $-\frac{B_{2k}}{2k}$.

3.1.3   The Zeta Function of the Laplacian on $\mathbb{S}^2$

Let A be the Laplacian on the sphere $\mathbb{S}^2$, the eigenvalues of A are the numbers $n(n+1)$ for $n = 0, 1, 2, \ldots$, each eigenvalue having multiplicity $(2n+1)$. Let ζ_A be the associated zeta function defined for $\text{Re}(s) > 1$ by

$$\zeta_A(s) = \sum_{n=1}^{+\infty}\frac{2n+1}{n^s(n+1)^s}$$

Analytic Continuation of the Zeta Function of the Laplacian

The analytic continuation of this function (Birmingham and Sen 1987) can be obtained by the use of an asymptotic expansion when $t \to 0+$ of the function

$$\theta_A : t \mapsto \sum_{n=1}^{+\infty} (2n + 1)e^{-n(n+1)t}$$

since θ_A is related to ζ_A by the Mellin transformation

$$\int_0^{+\infty} t^{s-1} \theta_A(t)dt = \zeta_A(s)\Gamma(s)$$

But a more simple method is to use the Ramanujan summation. Since for $Re(s) > 1$ we are in a case of convergence, we have

$$\zeta_A(s) = \sum_{n\geq 1}^{\mathcal{R}} \frac{2n + 1}{n^s(n + 1)^s} + \int_1^{+\infty} \frac{2x + 1}{x^s(x + 1)^s}dx$$

In this expression the sum is an entire function by the theorem of analyticity, and the integral is

$$\int_1^{+\infty} \frac{2x + 1}{x^s(x + 1)^s}dx = \int_1^{+\infty} (2x + 1)(x^2 + x)^{-s}dx = \frac{2^{1-s}}{s - 1}$$

Thus the analytic continuation of ζ_A on $\mathbb{C}\setminus\{1\}$ is simply given by

$$\zeta_A(s) = \sum_{n\geq 1}^{\mathcal{R}} \frac{2n + 1}{n^s(n + 1)^s} + \frac{2^{1-s}}{s - 1} \tag{3.2}$$

and the evaluation of $\zeta_A(-p)$ for $p = 0, -1, -2, \ldots$, is easily done by the Ramanujan summation:

$$\zeta_A(-p) = \sum_{n\geq 1}^{\mathcal{R}} (2n + 1)n^p(n + 1)^p - \frac{2^{p+1}}{p + 1}$$

We find for example

$$\zeta_A(0) = -\frac{2}{3}, \ \zeta_A(-1) = -\frac{1}{15}$$

Note that by the use of binomial expansion we obtain

$$\zeta_A(-p) = \sum_{k=0}^{p} C_p^k \left(2\zeta^{\mathcal{R}}(-p-k-1) + \zeta^{\mathcal{R}}(-p-k)\right) - \frac{2^{p+1}}{p+1}$$

and since $\zeta^{\mathcal{R}}(-q) = \zeta(-q) + \frac{1}{q+1}$ for $q \neq -1$, it follows that

$$\zeta_A(-p) = \sum_{k=0}^{p} C_p^k \left(2\zeta(-p-k-1) + \zeta(-p-k)\right)$$

$$+ \sum_{k=0}^{p} C_p^k \left(2\frac{1}{p+k+2} + \frac{1}{p+k+1}\right) - \frac{2^{p+1}}{p+1}$$

and by a combinatorial identity we have finally

$$\zeta_A(-p) = \sum_{k=0}^{p} C_p^k \left(2\zeta(-p-k-1) + \zeta(-p-k)\right)$$

The Determinant of the Laplacian on $\mathbb{S}^2$

The product

$$det(A) = \prod_{n \geq 1} n(n+1)$$

is clearly divergent and it is classically defined by a well-known procedure to define infinite divergent products. The *zeta-regularization* of a divergent product $\prod_{n \geq 1} a(n)$ is defined by

$$\overset{reg}{\prod_{n \geq 1}} a(n) = e^{-Z'_a(0)}$$

where Z_a is defined near 0 by the analytic continuation of the function

$$Z_a : s \mapsto \sum_{n \geq 1}^{+\infty} \frac{1}{(a(n))^s}$$

which is assumed to be defined and analytic for $\text{Re}(s) > \alpha$ for some $\alpha \in \mathbb{R}$.

In our case we have $a(n) = n(n+1)$, thus $Z_a = \zeta_A$ is given by (3.2) and we have

$$- \zeta_A'(s) = \sum_{n\geq 1}^{\mathcal{R}} \frac{2n+1}{n^s(n+1)^s} Log(n(n+1)) + \frac{2^{1-s}Log(2)}{s-1} + \frac{2^{1-s}}{(s-1)^2} \qquad (3.3)$$

Thus

$$-\zeta_A'(0) = \sum_{n\geq 1}^{\mathcal{R}} (2n+1)Log(n(n+1)) - 2Log(2) + 2$$

$$= 2 \sum_{n\geq 1}^{\mathcal{R}} nLog(n) + 2 \sum_{n\geq 1}^{\mathcal{R}} (n+1)Log(n+1)$$

$$+ \sum_{n\geq 1}^{\mathcal{R}} (Log(n) - Log(n+1)) - 2Log(2) + 2$$

and by the shift property, we get

$$-\zeta_A'(0) = 4 \sum_{n\geq 1}^{\mathcal{R}} nLog(n) + \frac{3}{2}$$

since $\sum_{n\geq 1}^{\mathcal{R}} nLog(n) = -\zeta'(-1) - \frac{1}{4}$ we get for the determinant of the Laplacian on $\mathbb{S}^2$ the value

$$e^{-\zeta_A'(0)} = e^{-4\zeta'(-1)+\frac{1}{2}}$$

3.1.4 More Zeta Series

The preceding technique can be applied to zeta series of type

$$Z(s) = \sum_{n=1}^{+\infty} \frac{P'(n)}{(P(n))^s}$$

where P is a polynomial such that $P(x) \neq 0$ for $Re(x) \geq 1$.
  We have for $Re(s) > 1$

$$\sum_{n=1}^{+\infty} \frac{P'(n)}{(P(n))^s} = \sum_{n\geq 1}^{\mathcal{R}} \frac{P'(n)}{(P(n))^s} + \int_1^{+\infty} \frac{P'(x)}{(P(x))^s} dx$$

Again the sum is an entire function by the theorem of analyticity, and the integral is

$$\int_1^{+\infty} \frac{P'(x)}{(P(x))^s} dx = \frac{P(1)^{1-s}}{s-1}$$

Thus the analytic continuation of Z is given by

$$Z(s) = \sum_{n\geq 1}^{\mathcal{R}} \frac{P'(n)}{(P(n))^s} + \frac{P(1)^{1-s}}{s-1}$$

Examples

(1) The simplest example is given by $P(x) = x + a$ with $Re(a) > 0$. We have for $Re(s) > 1$

$$Z(s) = \sum_{n=1}^{+\infty} \frac{1}{(n+a)^s} = \sum_{n\geq 1}^{\mathcal{R}} \frac{1}{(n+a)^s} + \frac{(a+1)^{1-s}}{s-1}$$

Since Z is related to the Hurwitz zeta function by $\zeta(s, a) = a^{-s} + Z(s)$, we obtain the analytic continuation of the Hurwitz zeta function

$$\zeta(s, a) = \frac{(a+1)^{1-s}}{s-1} + a^{-s} + \sum_{n\geq 1}^{\mathcal{R}} \frac{1}{(n+a)^s}$$

which easily gives for $s = -k = 0, -1, -2, \ldots$, the values

$$\zeta(-k, a) = a^k + \frac{(a+1)^{k+1}}{k+1} + \sum_{n\geq 1}^{\mathcal{R}} (n+a)^k$$

this last sum is given by (1.18) and (2.2) in terms of Bernoulli polynomials

$$\sum_{n\geq 1}^{\mathcal{R}} (n+a)^k = \frac{1 - B_{k+1}(a)}{k+1} - a^k + \int_1^{a+1} x^k dx$$

thus we obtain

$$\zeta(-k, a) = -\frac{B_{k+1}(a)}{k+1}$$

We have also

$$Z'(s) = -\sum_{n\geq 1}^{\mathcal{R}} \frac{Log(n+a)}{(n+a)^s} - \frac{(a+1)^{1-s}}{s-1}(Log(a+1) + \frac{1}{s-1})$$

The value $Z'(0)$ is easily evaluated since by (2.2) we have

$$\sum_{n\geq 1}^{\mathcal{R}} Log(n+a) = -Log(\Gamma(a+1)) + Log(\sqrt{2\pi}) - 1 + (a+1)Log(a+1) - a$$

thus

$$Z'(0) = Log(\Gamma(a+1)) - Log(\sqrt{2\pi})$$

This gives the regularized product

$$\prod_{n\geq 1}^{reg}(n+a) = e^{-Z'(0)} = \frac{\sqrt{2\pi}}{\Gamma(a+1)}$$

(2) Take for $Re(s) > 1$ the function

$$Z(s) = \sum_{n=1}^{+\infty} \frac{2n+1}{(n^2+n-1)^s}$$

Then the analytic continuation of Z is given by

$$Z(s) = \sum_{n\geq 1}^{\mathcal{R}} \frac{2n+1}{(n^2+n+1)^s} + \frac{1}{s-1}$$

Note that for the function defined for $Re(s) > 1/2$ by

$$Z_0(s) = \sum_{n=1}^{+\infty} \frac{1}{(n^2+n-1)^s}$$

the analytic continuation is not so simple. We have for $Re(s) = \sigma > 1/2$

$$Z(s) = \sum_{n\geq 1}^{\mathcal{R}} \frac{1}{(n^2+n-1)^s} + \int_1^{+\infty} \frac{1}{(x^2+x-1)^s}dx$$

Here again the sum is an entire function by the theorem of analyticity, and now the integral is

$$\int_1^{+\infty} \frac{1}{x^{2s}}(1 + \frac{x-1}{x^2})^{-s}dx = \sum_{k=0}^{+\infty} \frac{(-1)^k s \ldots (s+k-1)}{k!} \int_1^{+\infty} \frac{(x-1)^k}{x^{2s+2k}}dx$$

the interchange of $\sum_{k=0}^{+\infty}$ and $\int_1^{+\infty}$ is justified by

$$\sum_{k=0}^{+\infty} \frac{|s| \ldots (|s|+k-1)}{k!} \int_1^{+\infty} \frac{(x-1)^k}{x^{2\sigma+2k}}dx = \int_1^{+\infty} \frac{1}{x^{2\sigma}}(1-\frac{x-1}{x^2})^{-|s|}dx < +\infty$$

We verify that the integral is a function $s \mapsto Y(s)$ with the expansion

$$Y(s) = \frac{1}{2s-1} - s(\frac{1}{2s} - \frac{1}{2s+1})$$
$$+ \frac{s(s+1)}{2}(\frac{1}{2s+1} + \frac{1}{2s+3} - \frac{1}{s+1})$$
$$+ \ldots$$

And we deduce that Z is analytic near every negative integer $-p$, with

$$Z(-p) = \sum_{n\geq 1}^{\mathcal{R}}(n^2+n-1)^p + Y(-p)$$

For example, we have

$$Z(0) = \sum_{n\geq 1}^{\mathcal{R}} 1 + Y(0) = \frac{1}{2} - \frac{3}{2} = -1$$

$$Z(-1) = \sum_{n\geq 1}^{\mathcal{R}} (n^2+n-1) + Y(-1) = \frac{1}{4} + \frac{3}{4} = 1$$

$$Z(-2) = \sum_{n\geq 1}^{\mathcal{R}} (n^2+n-1)^2 + Y(-2) = \frac{1}{20} - \frac{21}{20} = -1$$

$$Z(-3) = \sum_{n\geq 1}^{\mathcal{R}} (n^2+n-1)^3 + Y(-3) = \frac{5}{56} + \frac{51}{56} = 1$$

$$Z(-4) = \sum_{n\geq 1}^{\mathcal{R}} (n^2+n-1)^4 + Y(-4) = \frac{5}{28} - \frac{33}{28} = -1$$

3.1.5 A Modified Zeta Function

We now study the function defined for $Re(s) > 1$ by

$$\zeta_1(s) = \sum_{n=1}^{+\infty} \frac{1}{1+n^s}$$

Then the analytic continuation of Z is given by

$$Z(s) = \sum_{n\geq 1}^{\mathcal{R}} \frac{1}{1+n^s} + \int_1^{+\infty} \frac{1}{1+x^s}\,dx$$

For the integral term we have

$$\int_1^{+\infty} \frac{1}{1+x^s}\,dx = \int_1^{+\infty} \frac{1}{x^s}\frac{1}{1+\frac{1}{x^s}}\,dx = \sum_{k=0}^{+\infty}(-1)^k \int_1^{+\infty} \frac{1}{x^{s+ks}}\,dx$$

the interchange of $\sum_{k=0}^{+\infty}$ and $\int_1^{+\infty}$ is justified by

$$\frac{1}{x^s}\frac{1}{1+\frac{1}{x^s}} = \sum_{k=0}^{N-1} \frac{(-1)^k}{x^{s+ks}} + \frac{(-1)^N}{x^{Ns}(1+x^s)}$$

and $\lim_{N\to+\infty} \int_1^{+\infty} \frac{1}{x^{Ns}(1+x^s)}\,dx = 0$. We obtain for the integral term the simple expression

$$\int_1^{+\infty} \frac{1}{1+x^s}\,dx = \sum_{k=0}^{+\infty}(-1)^k \frac{1}{(k+1)s-1}$$

Thus the analytic continuation of ζ_1 is defined on $\mathbb{C}\backslash\{1, \frac{1}{2}, \frac{1}{3}, \ldots, 0\}$ by

$$\zeta_1(s) = \sum_{n\geq 1}^{\mathcal{R}} \frac{1}{1+n^s} - \sum_{k=1}^{+\infty}(-1)^k \frac{1}{ks-1}$$

For $-s \in \mathbb{C}\setminus\{1, \frac{1}{2}, \frac{1}{3}, \ldots, 0\}$ we have

$$\zeta_1(-s) = \sum_{n\geq 1}^{\mathcal{R}} \frac{n^s}{1+n^s} + \sum_{k=1}^{+\infty} (-1)^k \frac{1}{ks+1}$$

$$= \sum_{n\geq 1}^{\mathcal{R}} 1 - \sum_{n\geq 1}^{\mathcal{R}} \frac{1}{1+n^s} + \sum_{k=1}^{+\infty} (-1)^k \frac{1}{ks+1}$$

$$= \frac{1}{2} - \zeta_1(s) - \sum_{k=1}^{+\infty} (-1)^k \frac{1}{ks-1} + \sum_{k=1}^{+\infty} (-1)^k \frac{1}{ks+1}$$

We deduce that for the function ζ_1 we have the simple functional equation

$$\zeta_1(s) + \zeta_1(-s) = \frac{1}{2} + 2 \sum_{k=1}^{+\infty} (-1)^{k-1} \frac{1}{k^2 s^2 - 1}$$

3.1.6 The Sums $\sum_{n\geq 1}^{\mathcal{R}} n^k \varphi_f(n)$

We examine the relation between the sums $\sum_{n\geq 1}^{\mathcal{R}} e^{-nz} \varphi_f(n)$ and $\sum_{n\geq 1}^{\mathcal{R}} e^{-nz} f(n)$.

Theorem 11 *Let f be a function of moderate growth and for $0 < z < \pi$ let*

$$L_f(x,z) = \int_1^x e^{-zt} f(t)dt$$

Then we have

$$\sum_{n\geq 1}^{\mathcal{R}} e^{-nz} \varphi_f(n) = \frac{1}{1-e^{-z}} \sum_{n\geq 1}^{\mathcal{R}} e^{-nz} f(n) - \frac{e^{-z}}{z} \sum_{n\geq 1}^{\mathcal{R}} f(n) - e^{-z} \sum_{n\geq 1}^{\mathcal{R}} e^{nz} L_f(n,z) \quad (3.4)$$

Proof We have (by Example 1 of Sect. 1.4.1)

$$R_{e^{-zx}}(x) = -e^{-zx} \frac{e^z}{1-e^z} - \frac{e^{-z}}{z} \quad \text{and} \quad \varphi_{e^{-zx}}(x) = \frac{e^{-zx}-1}{1-e^z}$$

Then by Theorem 3 we get

$$\sum_{n\geq 1}^{\mathcal{R}} e^{-nz} \varphi_f(n) = -\frac{e^{-z}}{z} \sum_{n\geq 1}^{\mathcal{R}} f(n) - \frac{e^z}{1-e^z} \sum_{n\geq 1}^{\mathcal{R}} e^{-nz} f(n)$$

$$- \frac{e^z}{1-e^z} \int_1^2 e^{-zx} R_f(x)dx$$

It remains to prove that

$$\frac{e^z}{1-e^z} \int_1^2 e^{-zt} R_f(t)dt = e^{-z} \sum_{n\geq 1}^{\mathcal{R}} e^{nz} \int_1^n e^{-zt}f(t)dt$$

Consider the function G defined by

$$G(x,z) = e^{zx} \int_1^x e^{-zt}f(t)dt$$

We can evaluate $\sum_{n\geq 1}^{\mathcal{R}} G(n,z)$ by observing that the function G is the solution of the differential equation

$$\partial_x G - zG = f \text{ with } G(1,z) = 0$$

By (2.7) the condition $G(1,z) = 0$ gives $R_{\partial_x G} = \partial_x R_G$ and we deduce that the function R_G is a solution of the differential equation

$$\partial_x R_G - zR_G = R_f$$

This gives

$$R_G(x,z) = Ke^{zx} + e^{zx} \int_1^x e^{-zt}R_f(t)dt$$

Using the condition $\int_1^2 R_G(x)dx = 0$ and integration by parts we get

$$K = -\frac{e^z}{e^z - 1} \int_1^2 e^{-zt}R_f(t)dt$$

then $R_G(1,z) = Ke^z$ gives

$$\sum_{n\geq 1}^{\mathcal{R}} e^{nz} \int_1^n e^{-zt}f(t)dt = \frac{e^{2z}}{1-e^z} \int_1^2 e^{-zt}R_f(t)dt$$

$\square$

Application: Evaluation of $\sum_{n\geq 1}^{\mathcal{R}} n^k \varphi_f(n)$

For a function f of moderate growth we define the sequence of functions $(F_k)_{k\geq 0}$ by $F_0 = f$ and for $k \geq 1$

$$F_k(x) = \int_1^x F_{k-1}(t)dt \text{ or } F_{k+1}(x) = \int_1^x \frac{(x-t)^k}{k!}f(t)dt$$

Then the Taylor expansion of the function $z \mapsto e^{zx} L_f(x, z)$ of the preceding theorem is

$$e^{zx} L_f(x, z) = e^{zx} \int_1^x e^{-zt} f(t) dt = \sum_{k \geq 0} z^k \int_1^x \frac{(x-t)^k}{k!} f(t) dt = \sum_{k \geq 0} F_{k+1}(x) z^k$$

Thus, by derivation of (3.4) with respect to z, we get a relation between the sums

$$\sum_{n \geq 1}^{\mathcal{R}} n^k \varphi_f(n), \ \sum_{n \geq 1}^{\mathcal{R}} n^k f(n) \text{ and } C_k = \sum_{n \geq 1}^{\mathcal{R}} F_k(n)$$

that is

$$\sum_{n \geq 1}^{\mathcal{R}} \varphi_f(n) = \frac{3}{2} \sum_{n \geq 1}^{\mathcal{R}} f(n) - \sum_{n \geq 1}^{\mathcal{R}} n f(n) - C_1$$

$$\sum_{n \geq 1}^{\mathcal{R}} n \varphi_f(n) = \frac{5}{12} \sum_{n \geq 1}^{\mathcal{R}} f(n) + \frac{1}{2} \sum_{n \geq 1}^{\mathcal{R}} n f(n) - \frac{1}{2} \sum_{n \geq 1}^{\mathcal{R}} n^2 f(n) - C_1 + C_2$$

$$\sum_{n \geq 1}^{\mathcal{R}} n^2 \varphi_f(n) = \frac{1}{3} \sum_{n \geq 1}^{\mathcal{R}} f(n) - \frac{1}{6} \sum_{n \geq 1}^{\mathcal{R}} n f(n) + \frac{1}{2} \sum_{n \geq 1}^{\mathcal{R}} n^2 f(n) - \frac{1}{3} \sum_{n \geq 1}^{\mathcal{R}} n^3 f(n)$$
$$- C_1 + 2C_2 - 2C_3$$

$$\cdots$$

Example For $f(x) = 1/x$ we get

$$\sum_{n \geq 1}^{\mathcal{R}} H_n = \frac{3}{2}\gamma - Log(\sqrt{2\pi}) + \frac{1}{2}$$

$$\sum_{n \geq 1}^{\mathcal{R}} n H_n = \frac{5}{12}\gamma - Log(\sqrt{2\pi}) - \zeta'(-1) + \frac{7}{8}$$

$$\sum_{n \geq 1}^{\mathcal{R}} n^2 H_n = \frac{1}{3}\gamma - Log(\sqrt{2\pi}) - 2\zeta'(-1) + \zeta'(-2) + \frac{17}{34}$$

More generally we have (Candelpergher et al. 2010)

$$\sum_{m \geq 1}^{\mathcal{R}} m^p H_m = \frac{1 - B_{p+1}}{p+1}\gamma + \sum_{k=1}^{p} (-1)^k C_p^k \zeta'(-k) - Log(\sqrt{2\pi}) + r_p \text{ with } r_p \in \mathbb{Q}$$

3.2   Integration with Respect to a Parameter

3.2.1   Interchanging $\sum_{n\geq 1}^{\mathcal{R}}$ and $\int_I$

Theorem 12 *Let $(x, u) \mapsto f(x, u)$ be a function defined for $Re(x) > a$, with $a < 1$, and $u \in I$ where I is an interval of $\mathbb{R}$. We suppose that*

(a) for all $Re(x) > a$ the function $u \mapsto f(x, u)$ is integrable on I
(b) there is $\alpha < \pi$ and a function $u \mapsto C(u)$ integrable on I such that

$$|f(x, u)| \leq C(u)e^{\alpha|x|} \text{ for } Re(x) > a \text{ and } u \in I$$

Then the function $u \mapsto \sum_{n\geq 1}^{\mathcal{R}} f(n, u)$ is integrable on I and we have

$$\int_I \sum_{n\geq 1}^{\mathcal{R}} f(n, u)du = \sum_{n\geq 1}^{\mathcal{R}} \int_I f(n, u)du$$

Proof By the integral formula (1.14) defining the Ramanujan summation we have

$$\sum_{n\geq 1}^{\mathcal{R}} f(n, u) = \frac{f(1, u)}{2} + i \int_0^{+\infty} \frac{f(1 + it, u) - f(1 - it, u)}{e^{2\pi t} - 1} dt$$

Since by hypothesis

$$\frac{|f(1 + it, u) - f(1 - it, u)|}{e^{2\pi t} - 1} \leq C(u) \, 2e^\alpha \frac{e^{\alpha t}}{e^{2\pi t} - 1}$$

we get

$$|\sum_{n\geq 1}^{\mathcal{R}} f(n, u)| \leq C(u) \left(\frac{1}{2}e^\alpha + \int_0^{+\infty} 2e^\alpha \frac{e^{\alpha t}}{e^{2\pi t} - 1} dt\right)$$

which proves that the function $u \mapsto \sum_{n\geq 1}^{\mathcal{R}} f(n, u)$ is integrable on I. Therefore

$$\int_I \sum_{n\geq 1}^{\mathcal{R}} f(n, u)du = \int_I \frac{f(1, u)}{2} du + i \int_I \int_0^{+\infty} \frac{f(1 + it, u) - f(1 - it, u)}{e^{2\pi t} - 1} dt du$$

It remains to prove that this last integral is

$$\int_0^{+\infty} \frac{\int_I f(1 + it, u)du - \int_I f(1 - it, u)du}{e^{2\pi t} - 1} dt$$

This is a consequence of the Fubini theorem since

$$\int_I \int_0^{+\infty} \frac{|f(1+it,u) - f(1-it,u)|}{e^{2\pi t} - 1} dt du \leq \int_I \int_0^{+\infty} C(u) \frac{2e^\alpha e^{\alpha t}}{e^{2\pi t} - 1} dt du < +\infty$$

$\square$

Example In Sect. 1.4.1 we have seen that

$$\sum_{n \geq 1}^{\mathcal{R}} \sin(\pi nt) = -\frac{\cos(\pi t)}{\pi t} + \frac{1}{2} \cot(\frac{\pi t}{2}) \text{ for } t \in [0, 1[$$

this gives

$$\int_0^{1/2} \sum_{n \geq 1}^{\mathcal{R}} t \sin(\pi nt) dt = -\frac{1}{\pi^2} + \frac{2}{\pi^2} \int_0^{\pi/4} t \cot(t) dt$$

By the preceding theorem this is equal to

$$\sum_{n \geq 1}^{\mathcal{R}} \int_0^{1/2} t \sin(\pi nt) dt = \sum_{n \geq 1}^{\mathcal{R}} (-\frac{\cos \frac{1}{2} \pi n}{2\pi n} + \frac{\sin(\frac{1}{2}\pi n)}{\pi^2 n^2})$$

and since we are in a case of convergence this last sum is

$$\frac{1}{4\pi} \sum_{n=1}^{\infty} \frac{(-1)^{n-1}}{n} + \frac{1}{\pi^2} \sum_{n=1}^{\infty} \frac{(-1)^{n-1}}{(2n-1)^2} + \int_1^{+\infty} \frac{\cos \frac{1}{2} \pi x}{2\pi x} dx - \int_1^{+\infty} \frac{\sin \frac{1}{2} \pi x}{\pi^2 x^2} dx$$

Finally using integration by parts we get

$$\int_0^{\pi/4} x \cot(x) dx = \frac{1}{2} G + \frac{1}{8} \pi Log(2) \text{ (}G \text{ is the Catalan's constant)}$$

And by the same type of calculation we get

$$\int_0^{\pi/4} x^2 \cot(x) dx = \frac{1}{4} \pi G + \frac{1}{32} \pi^2 Log(2) - \frac{35}{64} \zeta(3)$$

3.2.2   The Functional Equation for Zeta

By the integral formula (1.14) defining the Ramanujan summation we get for $u > 0$ the formula

$$\sum_{n\geq 1}^{\mathcal{R}} \frac{1}{n-1+u} = \frac{1}{2u} + i \int_0^{+\infty} \frac{\frac{1}{u+it} - \frac{1}{u-it}}{e^{2\pi t} - 1} dt$$

and by the use of the shift property we have

$$\sum_{n\geq 1}^{\mathcal{R}} \frac{1}{n+u} = Log(1 + \frac{1}{u}) - \frac{1}{2u} + 2 \int_0^{+\infty} \frac{1}{e^{2\pi t} - 1} \frac{t}{u^2 + t^2} dt$$

this gives the integral formula

$$\sum_{n\geq 1}^{\mathcal{R}} \frac{1}{n+u} = Log(1 + \frac{1}{u}) + 2 \int_0^{+\infty} (\frac{1}{e^{2\pi t} - 1} - \frac{1}{2\pi t}) \frac{t}{u^2 + t^2} dt$$

By the Mellin transform for $0 < Re(s) < 1$ of the two sides of this last relation we can obtain the functional equation of the Riemann zeta function:

(a) for the left side we can apply the preceding theorem since for $0 < a < 1$ we have for $Re(x) > a$

$$|\frac{u^{s-1}}{x+u}| \leq \frac{u^{Re(s)-1}}{a+u}$$

and the function $u \mapsto \frac{u^{Re(s)-1}}{a+u}$ is integrable on $]0, +\infty[$. Thus

$$\int_0^{+\infty} u^{s-1} \sum_{n\geq 1}^{\mathcal{R}} \frac{1}{n+u} du = \sum_{n\geq 1}^{\mathcal{R}} \int_0^{+\infty} u^{s-1} \frac{1}{n+u} du = \frac{\pi}{\sin \pi s} \sum_{n\geq 1}^{\mathcal{R}} n^{s-1}$$

(b) for the right side we get

$$\int_0^{+\infty} u^{s-1} Log(1 + \frac{1}{u}) du + 2 \int_0^{+\infty} u^{s-1} \int_0^{+\infty} (\frac{1}{e^{2\pi t} - 1} - \frac{1}{2\pi t}) \frac{t}{u^2 + t^2} dt \, du$$

which is

$$\frac{\pi}{s \sin \pi s} + 2 \int_0^{+\infty} (\frac{1}{e^{2\pi t} - 1} - \frac{1}{2\pi t}) \int_0^{+\infty} u^{s-1} \frac{t}{u^2 + t^2} du \, dt$$

From (a) and (b) and since $\sum_{n\geq1}^{\mathcal{R}} n^{s-1} = \zeta(1-s) + \frac{1}{s}$ we get for $0 < Re(s) < 1$

$$\frac{\pi}{\sin(\pi s)}\zeta(1-s) = 2\int_0^{+\infty}(\frac{1}{e^{2\pi t}-1} - \frac{1}{2\pi t})\int_0^{+\infty} u^{s-1}\frac{t}{u^2+t^2}du\,dt$$

Evaluation of this last integral (Titchmarsh and Heath-Brown 2007) gives the Riemann functional equation

$$\frac{\pi}{\sin \pi s}\zeta(1-s) = 2(2\pi)^{-s}\Gamma(s)\zeta(s)\frac{\pi/2}{\sin(\pi s/2)}$$

3.2.3   The Case of a Laplace Transform

Theorem 13 *Let $\hat{f}$ be a continuous function on $[0,+\infty[$ such that there is $a < 1$ and $C_a > 0$ with*

$$|\hat{f}(\xi)| \leq C_a e^{a\xi} \text{ for all } \xi \geq 0$$

and consider its Laplace transform f defined for $Re(x) > a$ by

$$f(x) = \int_0^{+\infty} e^{-x\xi}\,\hat{f}(\xi)d\xi$$

Then f is analytic for $Re(x) > a$ of moderate growth, and we have

$$\sum_{n\geq1}^{\mathcal{R}}f(n) = \int_0^{+\infty} e^{-\xi}(\frac{1}{1-e^{-\xi}} - \frac{1}{\xi})\hat{f}(\xi)d\xi\;. \tag{3.5}$$

Proof By hypothesis

$$|e^{-x\xi}\,\hat{f}(\xi)| \leq C_a e^{-(Re(x)-a)\xi}$$

and for $Re(x) > a$ we have

$$\int_0^{+\infty} e^{-(Re(x)-a)\xi}d\xi = \frac{1}{Re(x)-a}$$

This proves that the function f is analytic for $Re(x) > a$ and of moderate growth.

Then we apply the preceding theorem to the function $(x, \xi) \mapsto e^{-x\xi} \hat{f}(\xi)$, this gives

$$\sum_{n\geq 1}^{\mathcal{R}} \int_0^{+\infty} e^{-n\xi} \hat{f}(\xi)d\xi = \int_0^{+\infty} \sum_{n\geq 1}^{\mathcal{R}} e^{-n\xi} \hat{f}(\xi)d\xi$$

and since

$$\sum_{n\geq 1}^{\mathcal{R}} e^{-n\xi} = e^{-\xi}(\frac{1}{1-e^{-\xi}} - \frac{1}{\xi})$$

we get the result.   □

Remark With the same hypothesis of the preceding theorem we have an integral expression for the fractional remainder

$$R_f(x) = -\int_1^x f(t)dt + \int_0^{+\infty} e^{-x\xi}(\frac{1}{1-e^{-\xi}} - \frac{1}{\xi})\hat{f}(\xi)d\xi \tag{3.6}$$

Example Take $f(x) = \frac{1}{x^z}$ for $Re(z) > 0$ then

$$f(x) = \int_0^{+\infty} e^{-x\xi} \frac{\xi^{z-1}}{\Gamma(z)}d\xi$$

Thus for $Re(z) > 0$

$$\zeta(z) - \frac{1}{z-1} = \sum_{n\geq 1}^{\mathcal{R}} \frac{1}{n^z} = \int_0^{+\infty} e^{-\xi}(\frac{1}{1-e^{-\xi}} - \frac{1}{\xi}) \frac{\xi^{z-1}}{\Gamma(z)}d\xi .$$

and

$$\gamma = \sum_{n\geq 1}^{\mathcal{R}} \frac{1}{n} = \int_0^{+\infty} e^{-\xi}(\frac{1}{1-e^{-\xi}} - \frac{1}{\xi})d\xi .$$

Theorem 14 *Let f be a function of moderate growth such that for $Re(x) \geq 1$ we have*

$$f(x) = \sum_{k=1}^{+\infty} c_k \frac{1}{x^k}$$

where the power series $\sum_{k\geq 1} c_k x^k$ has a radius of convergence $\rho \geq 1$.

Then if $\rho > 1$, we have

$$\sum_{n\geq 1}^{\mathcal{R}} f(n) = c_1\gamma + \sum_{k=2}^{+\infty} c_k(\zeta(k) - \frac{1}{k-1}) \tag{3.7}$$

This result remains valid for $\rho = 1$ if this last series is convergent.

Proof We begin with the case $\rho > 1$.

For every $\rho - 1 < \varepsilon < \rho$ let $r = \frac{1}{\rho-\varepsilon} < 1$, by hypothesis there is $M > 0$ such that

$$|c_k| \leq Mr^k \text{ for every } k \geq 1$$

Thus the series $\sum_{k\geq 1} c_k \frac{\xi^{k-1}}{(k-1)!}$ is convergent for all $\xi \in \mathbb{C}$ and the function

$$\hat{f}(\xi) = \sum_{k=1}^{+\infty} c_k \frac{\xi^{k-1}}{(k-1)!}$$

is an entire function with $|\hat{f}(\xi)| \leq Ce^{r|\xi|}$. The function f is the Laplace transform of the entire function $\hat{f}$ because for $\text{Re}(x) > r$ we have

$$\int_0^{+\infty} \sum_{k=1}^{+\infty} c_k e^{-x\xi} \frac{\xi^{k-1}}{(k-1)!} d\xi = \sum_{k=1}^{+\infty} c_k \int_0^{+\infty} e^{-x\xi} \frac{\xi^{k-1}}{(k-1)!} d\xi = \sum_{k=1}^{+\infty} c_k \frac{1}{x^k} = f(x)$$

where the interchange of the signs $\int$ et $\sum$ is justified by

$$\int_0^{+\infty} \sum_{k=1}^{+\infty} |c_k| e^{-\text{Re}(x)\xi} \frac{\xi^{k-1}}{(k-1)!} d\xi \leq Mr \int_0^{+\infty} e^{-(\text{Re}(x)-r)\xi} d\xi < +\infty$$

Thus by Theorem 13 we get

$$\sum_{n\geq 1}^{\mathcal{R}} f(n) = \int_0^{+\infty} e^{-\xi}(\frac{1}{1-e^{-\xi}} - \frac{1}{\xi})(\sum_{k=1}^{+\infty} c_k \frac{\xi^{k-1}}{(k-1)!}) d\xi$$

Since

$$|e^{-\xi}(\frac{1}{1-e^{-\xi}} - \frac{1}{\xi})c_k \frac{\xi^{k-1}}{(k-1)!}| \leq e^{-\xi}(\frac{1}{1-e^{-\xi}} - \frac{1}{\xi})re^{r\xi} < re^{-(1-r)\xi}$$

we can again interchange $\int$ et $\sum$, and so we get

$$\sum_{n\geq 1}^{\mathcal{R}} f(n) = \sum_{k=1}^{\infty} c_k \int_0^{+\infty} e^{-\xi}(\frac{1}{1-e^{-\xi}} - \frac{1}{\xi}) \frac{\xi^{k-1}}{(k-1)!} d\xi .$$

and by the preceding example

$$\sum_{n\geq 1}^{\mathcal{R}} f(n) = c_1\gamma + \sum_{k=2}^{+\infty} c_k(\zeta(k) - \frac{1}{k-1})$$

It remains to extend this result to the case $\rho = 1$.

Let $0 < \alpha < 1$, if we apply the preceding case to the function $x \mapsto f(\frac{x}{\alpha})$ we get

$$\sum_{n\geq 1}^{\mathcal{R}} f(\frac{n}{\alpha}) = c_1\gamma\,\alpha + \sum_{k=2}^{+\infty} c_k(\zeta(k) - \frac{1}{k-1})\alpha^k$$

By the theorem of analyticity we have

$$\lim_{\alpha\to 1} \sum_{n\geq 1}^{\mathcal{R}} f(\frac{n}{\alpha}) = \sum_{n\geq 1}^{\mathcal{R}} f(n)$$

and by the Abel theorem on power series we have

$$\lim_{\alpha\to 1} \left(c_1\gamma\,\alpha + \sum_{k=2}^{+\infty} c_k(\zeta(k) - \frac{1}{k-1})\alpha^k\right) = c_1\gamma + \sum_{k=2}^{+\infty} c_k(\zeta(k) - \frac{1}{k-1})$$

if this last series is convergent. $\qquad\qquad\qquad\qquad\qquad\qquad\qquad\qquad\qquad$ $\square$

Remark Since

$$\sum_{n\geq 1}^{\mathcal{R}} \frac{1}{n} = \gamma \text{ and } \sum_{n\geq 1}^{\mathcal{R}} \frac{1}{n^k} = \zeta(k) - \frac{1}{k-1} \text{ for } k \neq 1$$

the preceding theorem can be stated in the form of the result of an interchange of the signs $\sum_{n\geq 1}^{\mathcal{R}}$ and $\sum_{k=1}^{+\infty}$:

$$\sum_{n\geq 1}^{\mathcal{R}} \sum_{k=1}^{+\infty} c_k \frac{1}{n^k} = \sum_{k=1}^{+\infty} c_k \sum_{n\geq 1}^{\mathcal{R}} \frac{1}{n^k}$$

Example Let $x \in \mathbb{R}$ and $f(x) = \sum_{n\geq 1}^{\mathcal{R}} \frac{e^{-\frac{x}{n}}}{n}$, we get by the preceding remark

$$f(x) = \sum_{k=0}^{+\infty} \frac{(-1)^k x^k}{k!} \sum_{n\geq 1}^{\mathcal{R}} \frac{1}{n^{k+1}} = \gamma + \sum_{k=1}^{+\infty} \frac{(-1)^k}{k!}(\zeta(k+1) - \frac{1}{k})x^k$$

Let us now see how this function is related to the heat equation in $\mathbb{R}^2$. Consider, for $X = (x_1, x_2) \in \mathbb{R}^2$ and $t \geq 0$, the function

$$U(X, t) = \sum_{n \geq 1}^{\mathcal{R}} \frac{1}{n + t} e^{-\frac{||X||^2}{4(n+t)}}$$

It is easily verified that this function is a solution of the heat equation

$$\partial_t U = \partial_{x_1 x_1}^2 U + \partial_{x_2 x_2}^2 U$$

Using the well known heat kernel we get

$$U(X, t) = \int_{\mathbb{R}^2} \frac{1}{4\pi t} e^{-\frac{||X-Y||^2}{4t}} U(y, 0) dY$$

Thus, using polar coordinates, we have for $t > 0$

$$\sum_{n \geq 1}^{\mathcal{R}} \frac{1}{(n + t)} e^{-\frac{r^2}{4(n+t)}} = e^{-\frac{r^2}{4t}} \frac{1}{2t} \int_0^{+\infty} I_0(\frac{r\rho}{2t}) \rho e^{-\frac{\rho^2}{4t}} (\sum_{n \geq 1}^{\mathcal{R}} \frac{1}{n} e^{-\frac{\rho^2}{4n}}) d\rho$$

where I_0 is the Bessel function

$$I_0(z) = \sum_{k=0}^{+\infty} \frac{1}{(k!)^2} (\frac{z}{2})^{2k}$$

With $x = \frac{r^2}{4}$ and $u = \frac{\rho^2}{4}$ we get

$$\sum_{n \geq 1}^{\mathcal{R}} \frac{1}{n + t} e^{-\frac{x}{n+t}} = e^{-\frac{x}{t}} \frac{1}{t} \int_0^{+\infty} I_0(\frac{2\sqrt{xu}}{t}) e^{-\frac{u}{t}} (\sum_{n \geq 1}^{\mathcal{R}} \frac{1}{n} e^{-\frac{u}{n}}) du$$

For $t = 1$, by the use of the shift property, we see that f verifies the integral equation

$$f(x) = e^{-x} - \int_1^2 \frac{1}{u} e^{-\frac{x}{u}} du + e^{-x} \int_0^{+\infty} I_0(2\sqrt{xu}) e^{-u} f(u) du$$

Note that

$$f(x) = \sum_{n \geq 1}^{\mathcal{R}} \frac{e^{-\frac{x}{n}} - 1}{n} + \gamma = \gamma + \sum_{n \geq 1}^{+\infty} \frac{e^{-\frac{x}{n}} - 1}{n} - \int_1^{+\infty} \frac{e^{-\frac{x}{v}} - 1}{v} dv$$

The function

$$g(x) = \gamma + \sum_{n \geq 1}^{+\infty} \frac{e^{-\frac{x}{n}} - 1}{n}$$

is the exponential generating function of the zeta values

$$g(x) = \gamma + \sum_{k=1}^{+\infty} \zeta(k+1)(-1)^k \frac{x^k}{k!}$$

It is easy to prove that the integral equation on f now gives a simpler integral equation for this generating function g, that is

$$g(x) = e^{-x} + e^{-x} \int_0^\infty e^{-u} I_0(2\sqrt{xu}) g(u) du$$

3.2.4   Series Involving Zeta Values

Consider an integer $m \geq 0$, by the shift property we have

$$\sum_{n \geq 1}^{\mathcal{R}} (n+1)^m Log(n+1) = \sum_{n \geq 1}^{\mathcal{R}} n^m Log(n) + \int_1^2 x^m Log(x) dx$$

On the other hand we have the expansion

$$(n+1)^m Log(n+1) = \sum_{j=0}^m C_m^j n^j Log(n) + \sum_{j=0}^m C_m^j n^j Log(1 + \frac{1}{n})$$

$$= \sum_{j=0}^m C_m^j n^j Log(n) + \sum_{j=0}^m C_m^j n^j \sum_{k=1}^{+\infty} \frac{(-1)^{k-1}}{k} \frac{1}{n^k}$$

Since

$$\sum_{n \geq 1}^{\mathcal{R}} n^j Log(n) = -\zeta'(-j) - \frac{1}{(j+1)^2}$$

we get the relation

$$-\sum_{j=0}^{m-1} C_m^j (\zeta'(-j)) + \sum_{j=0}^{m} C_m^j \sum_{k=1}^{+\infty} \frac{(-1)^{k-1}}{k} \zeta^{\mathcal{R}}(k-j) = \int_1^2 x^m Log(x) dx$$

$$+ \sum_{j=0}^{m-1} C_m^j (\frac{1}{j+1})^2$$

For example we get:
  with $m = 0$

$$\sum_{k=2}^{+\infty} \frac{(-1)^{k-1}}{k} \zeta(k) + \gamma = 0$$

with $m = 1$

$$Log(\sqrt{2\pi}) + \frac{1}{2}\gamma + \sum_{k=2}^{+\infty} \frac{(-1)^{k-1}}{k(k+1)} \zeta(k) = 1$$

with $m = 2$

$$Log(\sqrt{2\pi}) - 2\zeta'(-1) + \frac{1}{3}\gamma + 2\sum_{k=2}^{+\infty} \frac{(-1)^{k-1}}{k(k+1)(k+2)} \zeta(k) = \frac{4}{3}$$

with $m = 3$

$$Log(\sqrt{2\pi}) - 3\zeta'(-1) - 3\zeta'(-2) + \frac{1}{4}\gamma + 6\sum_{k=2}^{+\infty} \frac{(-1)^{k-1}}{k(k+1)(k+2)(k+3)} \zeta(k) = \frac{19}{12}$$

and more generally we have

$$-\ln(\sqrt{2\pi}) - \sum_{j=1}^{m-1} C_m^j \zeta'(-j) + \frac{1}{m+1}\gamma + m! \sum_{k=2}^{+\infty} \frac{(-1)^{k-1}}{k \dots (k+m)} \zeta(k) \in \mathbb{Q}$$

We can give another form of the preceding relation if we consider the function

$$\sum_{n\geq 1}^{\mathcal{R}} e^{-nz} Log(1 + \frac{1}{n}) = \sum_{k=0}^{+\infty} \frac{(-1)^k z^k}{k!} \sum_{j=1}^{+\infty} \frac{(-1)^{j-1}}{j} \zeta^{\mathcal{R}}(j-k)$$

if we use the formula we have proved in Sect. 2.1.2, that is

$$\sum_{n\geq1}^{\mathcal{R}} e^{-nz}Log(1+\frac{1}{n}) = (e^z - 1)\sum_{n\geq1}^{\mathcal{R}} e^{-nz}Log(n) + \int_0^1 Log(t+1)e^{-zt}dt$$

and (by the theorem of analyticity) the expansion

$$\sum_{n\geq1}^{\mathcal{R}} e^{-zn}Log(n) = \sum_{k=0}^{+\infty} \frac{(-1)^{k-1}}{k!}z^k(\zeta'(-k) + \frac{1}{(k+1)^2})$$

we get the relation

$$\sum_{k=0}^{+\infty} \frac{(-1)^kz^k}{k!} \sum_{j=1}^{+\infty} \frac{(-1)^{j-1}}{j}\zeta^{\mathcal{R}}(j-k) = (1-e^z)\sum_{k=0}^{+\infty} \frac{(-1)^kz^k}{k!}\zeta'(-k)$$

$$+(1-e^z)\sum_{k=0}^{+\infty} \frac{(-1)^kz^k}{k!}\frac{1}{(k+1)^2}$$

$$+\int_0^1 Log(t+1)e^{-zt}dt$$

Expanding in powers of z this gives an explicit evaluation of the sums

$$\sum_{j=2}^{+\infty} \frac{(-1)^j}{j+k}\zeta(j)$$

We get for example

$$\sum_{j=2}^{+\infty} \frac{(-1)^j}{j+1}\zeta(j) = -Log(\sqrt{2\pi}) + \frac{1}{2}\gamma + 1 \text{ (Singh and Verma 1983)}$$

$$\sum_{j=2}^{+\infty} \frac{(-1)^j}{j+2}\zeta(j) = -2\,\zeta'(-1) - Log(\sqrt{2\pi}) + \frac{1}{3}\gamma + \frac{2}{3}$$

$$\sum_{j=2}^{+\infty} \frac{(-1)^j}{j+3}\zeta(j) = 3\,\zeta'(-2) - 3\,\zeta'(-1) - Log(\sqrt{2\pi}) + \frac{1}{4}\gamma + \frac{7}{12}$$

$$\sum_{j=2}^{+\infty} \frac{(-1)^j}{j+4}\zeta(j) = -4\,\zeta'(-3) + 6\,\zeta'(-2) - 4\,\zeta'(-1) - Log(\sqrt{2\pi}) + \frac{1}{5}\gamma + \frac{47}{90}$$

Remark Note that a series like $\sum_{k \geq 2} \frac{\zeta(k)}{k}$ is divergent but we have

$$\sum_{k=2}^{+\infty} \frac{\zeta(k) - 1}{k} = \sum_{n=2}^{+\infty} (-Log(1 - \frac{1}{n}) - \frac{1}{n})$$

$$= \sum_{n=1}^{+\infty} (Log(n + 1) - Log(n) - \frac{1}{n + 1})$$

$$= \sum_{n \geq 1}^{\mathcal{R}} Log(n + 1) - \sum_{n \geq 1}^{\mathcal{R}} Log(n) - \sum_{n \geq 1}^{\mathcal{R}} \frac{1}{n + 1}$$

$$+ \int_{1}^{+\infty} (Log(x + 1) - Log(x) - \frac{1}{x + 1})$$

$$= 1 - \gamma$$

This technique can be applied to evaluate the sum $\sum_{k=1}^{+\infty} \frac{\zeta(2k)-1}{k+1}$. Since

$$\sum_{k=1}^{+\infty} \frac{\zeta(2k) - 1}{k + 1} = \sum_{n=1}^{+\infty} \sum_{k=1}^{+\infty} \frac{((n + 1)^{-2})^{k+1}}{k + 1} (n + 1)^2$$

we can write

$$\sum_{k=1}^{+\infty} \frac{\zeta(2k) - 1}{k + 1} = \sum_{n=1}^{+\infty} f(n)$$

where $f(n)$ is given by the function

$$f(x) = 2(x + 1)^2 Log(x + 1) - (x + 1)^2 (Log(x + 2) + Log(x)) - 1$$

Thus we have

$$\sum_{k=1}^{+\infty} \frac{\zeta(2k) - 1}{k + 1} = \sum_{n \geq 1}^{\mathcal{R}} f(n) + \int_{1}^{+\infty} f(x)dx = \frac{3}{2} - Log(\pi)$$

And, using the shift property, we deduce that

$$\sum_{k=1}^{+\infty} \frac{\zeta(2k) - 1}{k + 1} = \frac{3}{2} - Log(\pi)$$

We can apply the same method to get

$$\sum_{k=1}^{+\infty} \frac{\zeta(2k+1)-1}{k+1} = Log(2) - \gamma$$

3.3 Double Sums

3.3.1 Definitions and Properties

We study iterate Ramanujan summations

$$\sum_{n\geq 1}^{\mathcal{R}} \sum_{m\geq 1}^{\mathcal{R}} f(m,n)$$

Theorem 15 *Consider a function* $(x,y) \mapsto f(x,y)$ *analytic for* $Re(x) > 0$ *and* $Re(y) > 0$. *If there is* $C > 0$ *and* $\alpha < \pi$ *such that*

$$|f(x,y)| \leq Ce^{\alpha(|x|+|y|)} \tag{3.8}$$

then we have

$$\sum_{n\geq 1}^{\mathcal{R}} \sum_{m\geq 1}^{\mathcal{R}} f(m,n) = \sum_{m\geq 1}^{\mathcal{R}} \sum_{n\geq 1}^{\mathcal{R}} f(m,n)$$

Proof First we note that (3.8) implies that

$$x \mapsto f(x,y) \text{ in } \mathcal{O}^{\pi} \text{ for all } Re(y) > 0$$
$$y \mapsto f(x,y) \text{ in } \mathcal{O}^{\pi} \text{ for all } Re(x) > 0$$

For $Re(y) > 0$ consider the function $f_y : x \mapsto f(x,y)$ and its fractional remainder $R_{f_y} \in \mathcal{O}^{\pi}$. If we set $R(x,y) = R_{f_y}(x)$ then we have

$$R(x,y) - R(x+1,y) = f(x,y) \tag{3.9}$$

with $\int_1^2 R(x,y)dx = 0$. Thus for $Re(y) > 0$ we have

$$\sum_{m\geq 1}^{\mathcal{R}} f(m,y) = R(1,y) \tag{3.10}$$

This function R is given in Theorem 1 by the integral expression

$$R(x, y) = -\int_1^x f(t, y)dt + \frac{f(x, y)}{2} + i\int_0^{+\infty} \frac{f(x + it, y) - f(x - it, y)}{e^{2\pi t} - 1}dt$$

By this integral expression we see that the function R satisfies an inequality like (3.8) (with another constant C).

Thus for $Re(x) > 0$ the function $y \mapsto R(x, y)$ is in $\mathcal{O}^\pi$, thus by Theorem 1 we get a function $y \mapsto W(x, y)$ in $\mathcal{O}^\pi$ such that

$$W(x, y) - W(x, y + 1) = R(x, y) \tag{3.11}$$

and $\int_1^2 W(x, y)dy = 0$ for all $Re(x) > 0$.

This function also has the integral expression

$$W(x, y) = -\int_1^y R(x, t)dt + \frac{R(x, y)}{2} + i\int_0^{+\infty} \frac{R(x, y + it) - R(x, y - it)}{e^{2\pi t} - 1}dt$$

By definition of the Ramanujan summation we have for all $Re(x) > 0$

$$\sum_{n \geq 1}^{\mathcal{R}} R(x, n) = W(x, 1) \tag{3.12}$$

By (3.12) and (3.10) we get

$$W(1, 1) = \sum_{n \geq 1}^{\mathcal{R}} R(1, n) = \sum_{n \geq 1}^{\mathcal{R}} \sum_{m \geq 1}^{\mathcal{R}} f(m, n) \tag{3.13}$$

Now it remains to prove that $W(1, 1) = \sum_{m \geq 1}^{\mathcal{R}} \sum_{n \geq 1}^{\mathcal{R}} f(m, n)$.

By (3.9) and (3.11) we have the relation

$$W(x, y) - W(x + 1, y) - (W(x, y + 1) - W(x + 1, y + 1)) = f(x, y)$$

Thus if we set

$$T(x, y) = W(x, y) - W(x + 1, y)$$

and define for $Re(x) > 0$ the function f_x by $f_x(y) = f(x, y)$ then

$$T(x, y) - T(x, y + 1) = f_x(y)$$

Using the above integral expression of $W(x, y)$ we verify that W also satisfies an inequality like (3.8). Thus for $Re(x) > 0$ the function $y \mapsto T(x, y)$ is in $\mathcal{O}^\pi$. Since we have

$$\int_1^2 T(x, y)dy = \int_1^2 W(x, y)dy - \int_1^2 W(x + 1, y)dy = 0$$

then it follows that for $Re(x) > 0$ we have

$$R_{f_x}(y) = T(x, y)$$

We deduce that

$$\sum_{n\geq 1}^{\mathcal{R}} f(x, n) = T(x, 1) = W(x, 1) - W(x + 1, 1)$$

This gives for any integer $m \geq 1$

$$\sum_{n\geq 1}^{\mathcal{R}} f(m, n) = W(m, 1) - W(m + 1, 1)$$

Since the function $x \mapsto W(x, y)$ is in $\mathcal{O}^\pi$ for $Re(y) > 0$ then

$$\sum_{m\geq 1}^{\mathcal{R}} (W(m, 1) - W(m + 1, 1))$$

is well defined and we have by the shift property

$$\sum_{m\geq 1}^{\mathcal{R}} \sum_{n\geq 1}^{\mathcal{R}} f(m, n) = \sum_{m\geq 1}^{\mathcal{R}} W(m, 1) - \sum_{m\geq 1}^{\mathcal{R}} W(m + 1, 1) = W(1, 1) - \int_1^2 W(x, 1)dx$$

It remains to note that by (3.12) and Theorem 12 we have

$$\int_1^2 W(x, 1)dx = \int_1^2 \sum_{n\geq 1}^{\mathcal{R}} R(x, n)dx = \sum_{n\geq 1}^{\mathcal{R}} \int_1^2 R(x, n)dx = 0$$

$\square$

Remark Note that the sum $\sum_{k\geq 2}^{\mathcal{R}} \sum_{m+n=k} f(n + m)$ is well defined for $f \in \mathcal{O}^\pi$ since we have for $k \geq 2$

$$\sum_{m+n=k} f(n + m) = (k - 1)f(k)$$

Thus we have

$$\sum_{k\geq 2}^{\mathcal{R}} \sum_{m+n=k} f(n+m) = \sum_{k\geq 2}^{\mathcal{R}} (k-1)f(k) = \sum_{k\geq 1}^{\mathcal{R}} nf(n+1)$$

But we don't have equality between the sums

$$\sum_{m\geq 1}^{\mathcal{R}} \sum_{n\geq 1}^{\mathcal{R}} f(n+m) \text{ and } \sum_{k\geq 2}^{\mathcal{R}} \sum_{m+n=k} f(n+m).$$

Let us examine the relation between these sums. Let $F(x) = \int_1^x f(t)dt$ then by the shift property and Theorem 5 we have

$$\sum_{m\geq 1}^{\mathcal{R}} \sum_{n\geq 1}^{\mathcal{R}} f(n+m) = \frac{1}{2} \sum_{n\geq 1}^{\mathcal{R}} f(n) - \sum_{m\geq 1}^{\mathcal{R}} \varphi_f(m) + \sum_{m\geq 1}^{\mathcal{R}} \int_1^{m+1} f(x)dx$$

$$= \sum_{n\geq 1}^{\mathcal{R}} (n-1)f(n) + 2 \sum_{n\geq 1}^{\mathcal{R}} \int_1^n f(x)dx + \sum_{m\geq 1}^{\mathcal{R}} \int_m^{m+1} f(x)dx$$

$$= \sum_{n\geq 1}^{\mathcal{R}} (n-1)f(n) + 2 \sum_{n\geq 1}^{\mathcal{R}} F(n) + \int_1^2 F(y)dy$$

Next we observe that

$$\sum_{k\geq 2}^{\mathcal{R}} \sum_{m+n=k} f(n+m) = \sum_{n\geq 1}^{\mathcal{R}} nf(n+1)$$

$$= \sum_{n\geq 1}^{\mathcal{R}} (n-1)f(n) + \int_1^2 (x-1)f(x)dx$$

$$= \sum_{n\geq 1}^{\mathcal{R}} (n-1)f(n) + F(2) - \int_1^2 F(x)dx$$

Thus we obtain the following *formula of diagonal summation*

$$\sum_{m\geq 1}^{\mathcal{R}} \sum_{n\geq 1}^{\mathcal{R}} f(n+m) = \sum_{k\geq 2}^{\mathcal{R}} \sum_{m+n=k} f(n+m) + 2 \sum_{n\geq 1}^{\mathcal{R}} F(n) - F(2) + 2 \int_1^2 F(x)dx$$

or in another form

$$\sum_{m\geq 1}^{\mathcal{R}} \sum_{n\geq 1}^{\mathcal{R}} f(n+m) = \sum_{n\geq 1}^{\mathcal{R}} (n-1)f(n) + 2 \sum_{n\geq 1}^{\mathcal{R}} F(n) + \int_1^2 F(y)dy \qquad (3.14)$$

Example Consider the sum $\sum_{n\geq1}^{\mathcal{R}}\sum_{m\geq1}^{\mathcal{R}}\frac{m}{m+n}$. We have

$$\sum_{n\geq1}^{\mathcal{R}}\sum_{m\geq1}^{\mathcal{R}}\frac{m}{m+n}=\sum_{n\geq1}^{\mathcal{R}}\sum_{m\geq1}^{\mathcal{R}}\frac{m+n}{m+n}-\sum_{n\geq1}^{\mathcal{R}}\sum_{m\geq1}^{\mathcal{R}}\frac{n}{m+n}$$

$$=\frac{1}{4}-\sum_{n\geq1}^{\mathcal{R}}\sum_{m\geq1}^{\mathcal{R}}\frac{n}{m+n}$$

$$=\frac{1}{4}-\sum_{m\geq1}^{\mathcal{R}}\sum_{n\geq1}^{\mathcal{R}}\frac{n}{m+n}$$

thus we get

$$\sum_{n\geq1}^{\mathcal{R}}\sum_{m\geq1}^{\mathcal{R}}\frac{m}{m+n}=\frac{1}{8}$$

Since $\sum_{m\geq1}^{\mathcal{R}}\frac{1}{m+n}=\gamma-H_n+Log(n+1)$, we get

$$\sum_{n\geq1}^{\mathcal{R}}nH_n=\frac{5}{12}\gamma+\sum_{n\geq1}^{\mathcal{R}}nLog(n+1)-\frac{1}{8}$$

Note that

$$\sum_{m\geq1}^{\mathcal{R}}nLog(n+1)=\sum_{n\geq1}^{\mathcal{R}}(n+1)Log(n+1)-\sum_{n\geq1}^{\mathcal{R}}Log(n+1)$$

$$=-\zeta'(-1)-Log(\sqrt{2\pi})+1$$

thus we get another proof of the relation

$$\sum_{n\geq1}^{\mathcal{R}}nH_n=\frac{5}{12}\gamma-Log(\sqrt{2\pi})-\zeta'(-1)+\frac{7}{8}$$

More generally for a positive integer q we have

$$\sum_{k=0}^{2q}(-1)^kn^km^{2q-k}=\frac{m^{2q+1}+n^{2q+1}}{m+n}$$

and by the preceding theorem

$$\sum_{n\geq 1}^{\mathcal{R}}\sum_{m\geq 1}^{\mathcal{R}}\frac{m^{2q+1}+n^{2q+1}}{m+n} = \sum_{n\geq 1}^{\mathcal{R}}\sum_{m\geq 1}^{\mathcal{R}}\frac{m^{2q+1}}{m+n} + \sum_{n\geq 1}^{\mathcal{R}}\sum_{m\geq 1}^{\mathcal{R}}\frac{n^{2q+1}}{m+n}$$

$$= \sum_{n\geq 1}^{\mathcal{R}}\sum_{m\geq 1}^{\mathcal{R}}\frac{m^{2q+1}}{m+n} + \sum_{n\geq 1}^{\mathcal{R}}\sum_{m\geq 1}^{\mathcal{R}}\frac{n^{2q+1}}{m+n}$$

$$= 2\sum_{n\geq 1}^{\mathcal{R}}\sum_{m\geq 1}^{\mathcal{R}}\frac{n^{2q+1}}{m+n}$$

thus we get

$$\sum_{k=0}^{2q}(-1)^k\zeta^{\mathcal{R}}(-k)\zeta^{\mathcal{R}}(-2q+k) = 2\sum_{n\geq 1}^{\mathcal{R}}n^{2q+1}\sum_{m\geq 1}^{\mathcal{R}}\frac{1}{m+n}$$

then we have

$$\sum_{n\geq 1}^{\mathcal{R}}n^{2q+1}H_n = \sum_{n\geq 1}^{\mathcal{R}}n^{2q+1}Log(n+1) + \gamma\zeta^{\mathcal{R}}(-2q-1)$$

$$-\frac{1}{2}\sum_{k=0}^{2q}(-1)^k\zeta^{\mathcal{R}}(-k)\zeta^{\mathcal{R}}(-2q+k)$$

this gives $\sum_{n\geq 1}^{\mathcal{R}}n^{2q+1}H_n$ in terms of the values $\zeta'(-k)$ for $k = 0, 1, \ldots, 2q+1$.

3.3.2 Some Formulas for the Stieltjes Constant γ_1

The Stieltjes constant $\gamma_1 = \sum_{m\geq 1}^{\mathcal{R}}\frac{Log(m)}{m}$ is related to the sum $\sum_{m\geq 1}^{\mathcal{R}}\frac{Log(m+1)}{m}$. We have

$$\sum_{m\geq 1}^{\mathcal{R}}\frac{Log(m+1)}{m} = \sum_{m\geq 1}^{\mathcal{R}}\frac{Log(m+1)}{m+1} + \sum_{m\geq 1}^{\mathcal{R}}\frac{Log(m+1)}{m+1}\frac{1}{m}$$

$$= \sum_{m\geq 1}^{\mathcal{R}}\frac{Log(m)}{m} + \int_1^2\frac{Log(x)}{x}dx + \sum_{m\geq 1}^{\mathcal{R}}\frac{Log(m+1)}{m+1}\frac{1}{m}$$

$$= \sum_{m\geq 1}^{\mathcal{R}}\frac{Log(m)}{m} + \frac{1}{2}Log^2(2) + \sum_{m\geq 1}^{\mathcal{R}}\frac{Log(m+1)}{m+1}\frac{1}{m}$$

Since for the last series $\sum_{m\geq 1} \frac{Log(m+1)}{m+1} \frac{1}{m}$ we are in a case of convergence, then we have

$$\sum_{m\geq 1}^{\mathcal{R}} \frac{Log(m+1)}{m+1} \frac{1}{m} = \sum_{m=1}^{+\infty} \frac{Log(m+1)}{m(m+1)} - \int_1^{+\infty} \frac{Log(x+1)}{x(x+1)} dx$$

$$= \sum_{m=1}^{+\infty} \frac{Log(m+1)}{m(m+1)} - \frac{\pi^2}{12} - \frac{1}{2} Log^2(2)$$

Finally we get

$$\sum_{m\geq 1}^{\mathcal{R}} \frac{Log(m+1)}{m} = \gamma_1 - \frac{\pi^2}{12} + \sum_{m=1}^{+\infty} \frac{Log(m+1)}{m(m+1)} \qquad (3.15)$$

Proposition *We have the relation*

$$\sum_{m\geq 1}^{\mathcal{R}} \frac{Log(m+1)}{m} = \sum_{m\geq 1}^{\mathcal{R}} \frac{H_m}{m} - \frac{\gamma^2}{2} \qquad (3.16)$$

And the Stieltjes constant γ_1 is given by

$$\gamma_1 = \frac{\pi^2}{6} - \frac{1}{2} + \frac{1}{2} \int_1^2 (\psi(x))^2 dx - \sum_{m=1}^{+\infty} \frac{Log(m+1)}{m(m+1)} \qquad (3.17)$$

Proof We have by the linearity property

$$\sum_{m\geq 1}^{\mathcal{R}} \frac{1}{m} \sum_{n\geq 1}^{\mathcal{R}} \frac{1}{m+n} + \sum_{m\geq 1}^{\mathcal{R}} \sum_{n\geq 1}^{\mathcal{R}} \frac{1}{n} \frac{1}{m+n} = \sum_{m\geq 1}^{\mathcal{R}} \sum_{n\geq 1}^{\mathcal{R}} \frac{\frac{1}{m} + \frac{1}{n}}{m+n} = \sum_{m\geq 1}^{\mathcal{R}} \sum_{n\geq 1}^{\mathcal{R}} \frac{1}{mn} = \gamma^2$$

Interchange of m and n gives trivially

$$\sum_{m\geq 1}^{\mathcal{R}} \frac{1}{m} \sum_{n\geq 1}^{\mathcal{R}} \frac{1}{m+n} = \sum_{n\geq 1}^{\mathcal{R}} \frac{1}{n} \sum_{m\geq 1}^{\mathcal{R}} \frac{1}{n+m}$$

and by the preceding theorem

$$\sum_{n\geq 1}^{\mathcal{R}} \frac{1}{n} \sum_{m\geq 1}^{\mathcal{R}} \frac{1}{n+m} = \sum_{m\geq 1}^{\mathcal{R}} \sum_{n\geq 1}^{\mathcal{R}} \frac{1}{n} \frac{1}{n+m}$$

thus we get

$$\sum_{m\geq1}^{\mathcal{R}}\frac{1}{m}\sum_{n\geq1}^{\mathcal{R}}\frac{1}{m+n}=\frac{1}{2}\gamma^2 \tag{3.18}$$

Since $\sum_{n\geq1}^{\mathcal{R}}\frac{1}{m+n}=\gamma-H_m+Log(m+1)$ we have

$$\sum_{m\geq1}^{\mathcal{R}}\frac{1}{m}\sum_{n\geq1}^{\mathcal{R}}\frac{1}{m+n}=\sum_{m\geq1}^{\mathcal{R}}\frac{1}{m}(\gamma-H_m+Log(m+1))$$

that is

$$\sum_{m\geq1}^{\mathcal{R}}\frac{1}{m}\sum_{n\geq1}^{\mathcal{R}}\frac{1}{m+n}=\gamma^2-\sum_{m\geq1}^{\mathcal{R}}\frac{H_m}{m}+\sum_{m\geq1}^{\mathcal{R}}\frac{Log(m+1)}{m}$$

which gives by (3.18)

$$\sum_{m\geq1}^{\mathcal{R}}\frac{Log(m+1)}{m}=\sum_{m\geq1}^{\mathcal{R}}\frac{H_m}{m}-\frac{\gamma^2}{2}$$

This last relation gives by (3.15)

$$\gamma_1=\sum_{m\geq1}^{\mathcal{R}}\frac{H_m}{m}-\frac{\gamma^2}{2}+\frac{\pi^2}{12}-\sum_{m=1}^{+\infty}\frac{Log(m+1)}{m(m+1)} \tag{3.19}$$

The conclusion is obtained by the use of (2.6) that is

$$\sum_{n\geq1}^{\mathcal{R}}\frac{H_n}{n}=\frac{1}{2}(\zeta(2)-1+\gamma^2)+\frac{1}{2}\int_1^2(\psi(t))^2dt$$

$\square$

Remark Note also that

$$\sum_{n\geq1}^{\mathcal{R}}\frac{Log(n+1)}{n}=\sum_{n\geq1}^{\mathcal{R}}\int_0^1\frac{1}{1+nx}dx$$

By Theorem 12 we have

$$\sum_{n\geq1}^{\mathcal{R}}\frac{Log(n+1)}{n}=\int_0^1\sum_{n\geq1}^{\mathcal{R}}\frac{1}{1+nx}dx$$

and by (1.34)

$$\sum_{n\geq 1}^{\mathcal{R}} \frac{1}{1+nx} dx = \frac{Log(\frac{1}{x}+1) - \psi(\frac{1}{x}+1)}{x}$$

therefore we have

$$\sum_{n\geq 1}^{\mathcal{R}} \frac{Log(n+1)}{n} = \int_1^{+\infty} \frac{Log(x+1) - \psi(x+1)}{x} dx$$

By (3.15) this gives

$$\gamma_1 = \frac{\pi^2}{12} - \sum_{m=1}^{+\infty} \frac{Log(m+1)}{m(m+1)} + \int_1^{+\infty} \frac{Log(x+1) - \psi(x+1)}{x} dx \qquad (3.20)$$

There are similar integral expressions for the sums $\sum_{n\geq 1}^{+\infty} \frac{Log(n+1)}{n^p}$ with an integer $p > 1$. Using the finite Taylor expansion we get

$$Log(x+1) = \sum_{k=1}^{p-2} \frac{(-1)^{k-1}}{k} x^k + (-1)^p x^{p-1} \int_1^{+\infty} \frac{1}{t^{p-1}} \frac{1}{t+x} dx$$

which gives for every integer $n \geq 1$

$$\frac{Log(n+1)}{n^p} = \sum_{k=1}^{p-2} \frac{(-1)^{k-1}}{k} \frac{1}{n^{p-k}} + (-1)^p \int_1^{+\infty} \frac{1}{t^{p-1}} \frac{1}{n(t+n)} dt$$

and by summation we get

$$\sum_{n=1}^{+\infty} \frac{Log(n+1)}{n^p} = \sum_{k=1}^{p-2} \frac{(-1)^k}{k} \zeta(p-k) + \frac{(-1)^p}{p-1} \gamma + (-1)^p \int_1^{+\infty} \frac{\psi(x+1)}{x^p} dx$$

Proposition *We also have the integral expressions*

$$\gamma_1 = \frac{\pi^2}{6} - \frac{1}{2} + \frac{1}{2} \int_0^1 (\psi(x+1))^2 dx - \int_0^1 \frac{\psi(x+1) + \gamma}{x} dx \qquad (3.21)$$

and

$$\gamma_1 = \frac{\pi^2}{6} - \left(\int_0^1 \frac{\psi(x+1) + \gamma}{x} dx + \int_1^{+\infty} \frac{\Psi(x+1) - Log(x)}{x} dx \right) \qquad (3.22)$$

Proof We can obtain another formula for γ_1 by using the relation

$$\sum_{m\geq 1}^{\mathcal{R}} \frac{1}{m+n-1} = \gamma - H_n + \frac{1}{n} + Log(n)$$

which gives

$$\sum_{n\geq 1}^{\mathcal{R}} \frac{Log(n)}{n} = \sum_{n\geq 1}^{\mathcal{R}}\sum_{m\geq 1}^{\mathcal{R}} \frac{1}{n}\frac{1}{m+n-1} - \gamma^2 + \sum_{n\geq 1}^{\mathcal{R}} \frac{H_n}{n} - \zeta(2) + 1$$

By the preceding theorem this last double sum is also

$$\sum_{n\geq 1}^{\mathcal{R}}\sum_{m\geq 1}^{\mathcal{R}} \frac{1}{n}\frac{1}{m+n-1} = \sum_{m\geq 1}^{\mathcal{R}}\sum_{n\geq 1}^{\mathcal{R}} \frac{1}{n}\frac{1}{m+n-1} = \sum_{n\geq 1}^{\mathcal{R}}\sum_{m\geq 1}^{\mathcal{R}} \frac{1}{m}\frac{1}{m+n-1}$$

thus we have

$$\gamma_1 = \sum_{n\geq 1}^{\mathcal{R}} f(n) - \gamma^2 + \sum_{n\geq 1}^{\mathcal{R}} \frac{H_n}{n} - \zeta(2) + 1$$

with

$$f(n) = \sum_{m\geq 1}^{\mathcal{R}} \frac{1}{m}\frac{1}{m+n-1}$$

For $n > 1$ we have by (1.34)

$$f(n) = \frac{1}{n-1}\Big(\sum_{m\geq 1}^{\mathcal{R}} \frac{1}{m} - \sum_{m\geq 1}^{\mathcal{R}} \frac{1}{m+n-1}\Big) = \frac{1}{n-1}(\gamma + \psi(n) - Log(n))$$

and for $n = 1$ we have $f(n) = \zeta(2) - 1$. Thus we see that the function f is given by

$$f(x) = \frac{\gamma + \psi(x) - Log(x)}{x-1}$$

$$f(1) = \zeta(2) - 1$$

Using the shift property we have

$$\sum_{n\geq 1}^{\mathcal{R}} f(n) = \sum_{n\geq 1}^{\mathcal{R}} \frac{H_n}{n} - \sum_{n\geq 1}^{\mathcal{R}} \frac{Log(n+1)}{n} + \zeta(2) - 1 - \int_1^2 \frac{\gamma + \psi(x) - Log(x)}{x-1} dx$$

which is by (3.16)

$$\sum_{n\geq 1}^{\mathcal{R}} f(n) = \frac{1}{2}\gamma^2 + \frac{\pi^2}{4} - 1 - \int_0^1 \frac{\psi(x+1)+\gamma}{x} dx$$

this gives

$$\gamma_1 = \sum_{n\geq 1}^{\mathcal{R}} \frac{H_n}{n} - \frac{1}{2}\gamma^2 + \frac{\pi^2}{12} - \int_0^1 \frac{\psi(x+1)+\gamma}{x} dx \qquad (3.23)$$

and (3.21) is obtained by using (2.6).

With the preceding results (3.17) and (3.21) we get

$$\int_0^1 \frac{\gamma + \psi(x+1)}{x} dx = \sum_{m=1}^{+\infty} \frac{Log(m+1)}{m(m+1)}$$

and by (3.20) we have

$$\gamma_1 = \frac{\pi^2}{12} - \int_1^{+\infty} \frac{\psi(x+1)+\gamma}{x} dx - \int_1^{+\infty} \frac{\psi(x+1) - Log(x+1)}{x} dx$$

this is (3.22) since

$$\int_1^{+\infty} \frac{Log(x+1) - Log(x)}{x} dx = \int_1^{+\infty} \frac{Log(\frac{1}{x}+1)}{x} dx = \frac{\pi^2}{12}$$

$\square$

Remark The series $\sum_{n\geq 1} \frac{1}{n}(H_n - Log(n) - \gamma)$ is convergent and we have

$$\sum_{n=1}^{+\infty} \frac{H_n - Log(n) - \gamma}{n} = \sum_{n=1}^{\mathcal{R}} \frac{H_n - Log(n) - \gamma}{n} + \int_1^{+\infty} \frac{\psi(x+1) - Log(x)}{x} dx$$

Since by (3.19) and (3.20) we have

$$\sum_{n\geq 1}^{\mathcal{R}} \frac{H_n}{n} = \frac{1}{2}\gamma^2 + \frac{\pi^2}{12} - \int_1^{+\infty} \frac{\psi(x+1) - Log(x)}{x} dx$$

this gives the following result (Furdui 2012)

$$\sum_{n=1}^{+\infty} \frac{1}{n}(H_n - Log(n) - \gamma) = -\frac{1}{2}\gamma^2 + \frac{\pi^2}{12} - \gamma_1$$

3.3.3   A Simple Proof of a Formula of Ramanujan

Now we use double sums to give a new proof of the Ramanujan's relation (Grosswald 1972): for $t > 0$ we have

$$\frac{1}{t}\sum_{m=1}^{+\infty}\frac{m\pi}{e^{2m\pi/t}-1}+t\sum_{m=1}^{+\infty}\frac{m\pi}{e^{2m\pi t}-1}=(t+\frac{1}{t})\frac{\pi}{24}-\frac{1}{4} \tag{3.24}$$

Let us start with the well known identities

$$\sum_{n=1}^{+\infty}\frac{x^2}{x^2+n^2}=x\frac{\pi}{2}+\frac{x\pi}{e^{2\pi x}-1}-\frac{1}{2}$$

$$\int_{1}^{+\infty}\frac{x^2}{x^2+t^2}dt=x\frac{\pi}{2}-x\,\arctan(\frac{1}{x})$$

We deduce that

$$\sum_{n\geq1}^{\mathcal{R}}\frac{x^2}{x^2+n^2}=\frac{x\pi}{e^{2\pi x}-1}-\frac{1}{2}+x\,\arctan(\frac{1}{x})$$

Let $t > 0$, using the preceding relation with $x = m/t$ we get

$$\sum_{m\geq1}^{\mathcal{R}}\sum_{n\geq1}^{\mathcal{R}}\frac{m^2}{m^2+n^2t^2}=\frac{1}{t}\sum_{m\geq1}^{\mathcal{R}}\frac{m\pi}{e^{2m\pi/t}-1}-\frac{1}{4}+\frac{1}{t}\sum_{m\geq1}^{\mathcal{R}}m\arctan(\frac{t}{m}) \tag{3.25}$$

Consider for $t > 0$ the functions

$$f(t)=\sum_{m\geq1}^{\mathcal{R}}\sum_{n\geq1}^{\mathcal{R}}\frac{m^2}{m^2+n^2t^2}$$

$$S(t)=\frac{1}{t}\sum_{m=1}^{+\infty}\frac{m\pi}{e^{2m\pi/t}-1}$$

$$A(t)=\frac{1}{t}\sum_{m\geq1}^{\mathcal{R}}m\arctan(\frac{t}{m})$$

Since we have

$$\frac{1}{t}\sum_{m\geq1}^{\mathcal{R}}\frac{m\pi}{e^{2m\pi/t}-1}=S(t)-\frac{t}{4\pi}\int_{\frac{2\pi}{t}}^{+\infty}\frac{x}{e^x-1}dx$$

then by (3.25) we have the relation

$$S(t) = f(t) - A(t) + \frac{1}{4} + \frac{t}{4\pi} \int_{\frac{2\pi}{t}}^{+\infty} \frac{x}{e^x - 1} dx \qquad (3.26)$$

The relation (3.24) is simply $S(t) + S(\frac{1}{t}) = (t + \frac{1}{t})\frac{\pi}{24} - \frac{1}{4}$.
To prove this relation we use the following expression of $A(t)$.

Lemma *We have*

$$A(t) = \frac{1}{2}\left((t + \frac{1}{t}) \arctan(t) + 1\right) - \frac{1}{4\pi t} \int_0^{2\pi t} \frac{x}{e^x - 1} dx - \frac{\pi t}{4}$$

From (3.26) we deduce that

$$S(t) = f(t) + g(t) + \frac{1}{4} + \frac{\pi t}{4} - \frac{1}{2}\left((t + \frac{1}{t}) \arctan(t) + 1\right) \qquad (3.27)$$

where

$$g(t) = \frac{t}{4\pi} \int_{\frac{2\pi}{t}}^{+\infty} \frac{x}{e^x - 1} dx + \frac{1}{4\pi t} \int_0^{2\pi t} \frac{x}{e^x - 1} dx$$

For this last function g the equation

$$g(t) + g(\frac{1}{t}) = \left(\frac{t}{4\pi} + \frac{1}{4\pi t}\right) \int_0^{+\infty} \frac{x}{e^x - 1} dx$$

is immediately verified. And since $\int_0^{+\infty} \frac{x}{e^x - 1} dx = \frac{\pi^2}{6}$ we get

$$g(t) + g(\frac{1}{t}) = \frac{\pi t}{24} + \frac{\pi}{24t} \qquad (3.28)$$

It remains to find a similar relation for the function f. By Theorem 15 we get

$$f(\frac{1}{t}) = \sum_{m \geq 1}^{\mathcal{R}} \sum_{n \geq 1}^{\mathcal{R}} \frac{m^2 t^2}{m^2 t^2 + n^2}$$

$$= \sum_{n \geq 1}^{\mathcal{R}} \sum_{m \geq 1}^{\mathcal{R}} \frac{m^2 t^2}{m^2 t^2 + n^2}$$

$$= \sum_{m \geq 1}^{\mathcal{R}} \sum_{n \geq 1}^{\mathcal{R}} \frac{n^2 t^2}{n^2 t^2 + m^2}$$

We have

$$f(t) + f(\frac{1}{t}) = \sum_{m\geq 1}^{\mathcal{R}}\sum_{n\geq 1}^{\mathcal{R}} \frac{m^2 + n^2 t^2}{n^2 t^2 + m^2} = \sum_{m\geq 1}^{\mathcal{R}}\sum_{n\geq 1}^{\mathcal{R}} 1 = \sum_{m\geq 1}^{\mathcal{R}} \frac{1}{2} = \frac{1}{4}$$

Thus we obtain

$$f(t) + f(\frac{1}{t}) = \frac{1}{4} \tag{3.29}$$

Finally by (3.27), (3.28), and (3.29) we have

$$S(t) + S(\frac{1}{t}) = \frac{1}{4} + \frac{\pi t}{24} + \frac{\pi}{24t}$$

$$+ \frac{1}{4} + \frac{\pi t}{4} - \frac{1}{2}((t + \frac{1}{t})\arctan(t) + 1)$$

$$+ \frac{1}{4} + \frac{\pi}{4t} - \frac{1}{2}((t + \frac{1}{t})\arctan(\frac{1}{t}) + 1)$$

since $\arctan(t) + \arctan(\frac{1}{t}) = \frac{\pi}{2}$ this proves (3.16).

Proof of the Lemma Expanding the function arctan and by using Theorem 14 we get

$$A(t) = \sum_{m\geq 1}^{\mathcal{R}}\sum_{k=0}^{+\infty} \frac{(-1)^k}{2k+1}t^{2k}\frac{1}{m^{2k}} = \sum_{k=0}^{+\infty} \frac{(-1)^k t^{2k}}{2k+1}(\zeta(2k) - \frac{1}{2k-1})$$

Since $\zeta(2k) = \frac{(-1)^{k-1}}{2}\frac{(2\pi)^{2k}}{(2k)!}B_{2k}$ we have

$$\sum_{k=0}^{+\infty} \frac{(-1)^k t^{2k}}{2k+1}\zeta(2k) = -\frac{1}{2}\sum_{k=1}^{+\infty} \frac{B_{2k}}{2k+1}\frac{(2\pi t)^{2k}}{(2k)!} = -\frac{1}{4\pi t}\int_0^{2\pi t} \frac{x}{e^x - 1}dx - \frac{\pi t}{4}$$

and it is immediately verified that

$$-\sum_{k=0}^{+\infty} \frac{(-1)^k t^{2k}}{(2k-1)(2k+1)} = \frac{1}{2}((t + \frac{1}{t})\arctan(t) + 1)$$

Thus we get

$$A(t) = \frac{1}{2}((t + \frac{1}{t})\arctan(t) + 1) - \frac{1}{4\pi t}\int_0^{2\pi t} \frac{x}{e^x - 1}dx - \frac{\pi t}{4}$$

□

3.3.4   The Functional Relation for Eisenstein Function G_2

Let k be an integer > 1 and let G_{2k} be the Eisenstein function defined for $Im(z) > 0$ by

$$G_{2k}(z) = \sum_n \sum_m \frac{1}{(m + nz)^{2k}}$$

where this double sum is restricted to $(m, n) \in \mathbb{Z}\backslash\{(0, 0)\}$.
  It is well known that G_{2k} satisfies the modular relation

$$G_{2k}\left(-\frac{1}{z}\right) = z^{2k} G_{2k}(z)$$

For $k = 1$ we consider for $Im(z) > 0$ the Eisenstein function

$$G_2(z) = \sum_n \sum_m \frac{1}{(m + nz)^2}$$

where this double sum is restricted to $(m, n) \in \mathbb{Z}\backslash\{(0, 0)\}$, more precisely

$$G_2(z) = \sum_{n=-\infty}^{+\infty} \left(\sum_{\substack{m=-\infty \\ m \neq 0 \text{ if } n=0}}^{+\infty} \frac{1}{(m + nz)^2} \right)$$

Since we don't have absolute summability we cannot interchange $\sum_n$ and $\sum_m$ and we have only

$$G_2\left(-\frac{1}{z}\right) = z^2 \sum_m \sum_n \frac{1}{(m + nz)^2}$$

But it is well known (Freitag and Busam 2009) that this function G_2 satisfies the non trivial relation

$$G_2\left(-\frac{1}{z}\right) = z^2 G_2(z) - 2i\pi z \tag{3.30}$$

We now give a simple proof of this relation thanks to the Ramanujan summation.
  The relation (3.30) is equivalent to

$$\sum_m \sum_n \frac{1}{(m + nz)^2} = \sum_n \sum_m \frac{1}{(m + nz)^2} - \frac{2i\pi}{z}$$

where the sums are restricted to $(m, n) \neq (0, 0)$.

These sums can be written in another form, we have

$$\sum_n \sum_m \frac{1}{(m+nz)^2} = 2\sum_{n=1}^{+\infty}\sum_{m=1}^{+\infty}(\frac{1}{(m+nz)^2} + \frac{1}{(m-nz)^2}) + 2(1+\frac{1}{z^2})\zeta(2)$$

Thus if we set

$$A(z) = \sum_{n=1}^{+\infty}\sum_{m=1}^{+\infty}(\frac{1}{(m+nz)^2} + \frac{1}{(m-nz)^2})$$

and

$$B(z) = \sum_{m=1}^{+\infty}\sum_{n=1}^{+\infty}(\frac{1}{(m+nz)^2} + \frac{1}{(m-nz)^2})$$

we see that the relation (3.30) is equivalent to

$$A(z) - B(z) = \frac{i\pi}{z}$$

To compare $A(z)$ and $B(z)$ we can use the Ramanujan summation, since by Theorem 15 we have the commutation relation

$$\sum_{n\geq1}^{\mathcal{R}}\sum_{m\geq1}^{\mathcal{R}}(\frac{1}{(m+nz)^2} + \frac{1}{(m-nz)^2}) = \sum_{m\geq1}^{\mathcal{R}}\sum_{n\geq1}^{\mathcal{R}}(\frac{1}{(m+nz)^2} + \frac{1}{(m-nz)^2})$$

Next we use the relation between the Ramanujan summation and usual summation to transform this relation into a relation between the usual sums. Let us start with $A(z)$, we have

$$\sum_{m=1}^{+\infty}(\frac{1}{(m+nz)^2} + \frac{1}{(m-nz)^2}) = \sum_{m\geq1}^{\mathcal{R}}(\frac{1}{(m+nz)^2} + \frac{1}{(m-nz)^2})$$
$$+ \frac{1}{1+nz} + \frac{1}{1-nz}$$

this gives

$$\sum_{n=1}^{+\infty}\sum_{m=1}^{+\infty}(\frac{1}{(m+nz)^2} + \frac{1}{(m-nz)^2}) = \sum_{n=1}^{+\infty}\sum_{m\geq1}^{\mathcal{R}}(\frac{1}{(m+nz)^2} + \frac{1}{(m-nz)^2})$$
$$+ \sum_{n=1}^{+\infty}(\frac{1}{1+nz} + \frac{1}{1-nz})$$

Using again the relation between the Ramanujan summation and the usual Cauchy summation we obtain

$$\sum_{n=1}^{+\infty}\sum_{m\geq1}^{\mathcal{R}}(\frac{1}{(m+nz)^2}+\frac{1}{(m-nz)^2}) = \sum_{n\geq1}^{\mathcal{R}}\sum_{m\geq1}^{\mathcal{R}}(\frac{1}{(m+nz)^2}+\frac{1}{(m-nz)^2})$$

$$+ \int_1^{+\infty}\sum_{m\geq1}^{\mathcal{R}}(\frac{1}{(m+tz)^2}+\frac{1}{(m-tz)^2})dt$$

Interchanging Ramanujan summation and integration we get

$$\int_1^{+\infty}\sum_{m\geq1}^{\mathcal{R}}(\frac{1}{(m+tz)^2}+\frac{1}{(m-tz)^2})dt = \frac{1}{z}\sum_{m\geq1}^{\mathcal{R}}(\frac{1}{m+z}-\frac{1}{m-z})$$

Therefore we finally obtain

$$A(z) = \sum_{n\geq1}^{\mathcal{R}}\sum_{m\geq1}^{\mathcal{R}}(\frac{1}{(m+nz)^2}+\frac{1}{(m-nz)^2})$$

$$+ \sum_{n=1}^{+\infty}(\frac{1}{1+nz}+\frac{1}{1-nz})+\frac{1}{z}\sum_{m\geq1}^{\mathcal{R}}(\frac{1}{m+z}-\frac{1}{m-z})$$

The same calculation with $\sum_m\sum_n\frac{1}{(m+nz)^2}$ gives

$$B(z) = \sum_{m\geq1}^{\mathcal{R}}\sum_{n\geq1}^{\mathcal{R}}(\frac{1}{(m+nz)^2}+\frac{1}{(m-nz)^2})$$

$$+ \sum_{n\geq1}^{\mathcal{R}}(\frac{1}{1+nz}+\frac{1}{1-nz})+\frac{1}{z}\sum_{m=1}^{+\infty}(\frac{1}{m+z}-\frac{1}{m-z})$$

By Theorem 15 the double Ramanujan summations in the last formulas for $A(z)$ and $B(z)$ are the same, so we get

$$A(z) - B(z) = \int_1^{+\infty}(\frac{1}{1+tz}+\frac{1}{1-tz})dt - \frac{1}{z}\int_1^{+\infty}(\frac{1}{t+z}-\frac{1}{t-z})dt$$

and finally

$$A(z) - B(z) = \frac{1}{z}\lim_{T\to+\infty}(Log(1+Tz)-Log(1-Tz)) = \frac{i\pi}{z}$$

Chapter 4
Transformation Formulas

In this chapter we study some transformations that give interesting relations between the Ramanujan summation and other summations. In the first section we examine the Borel summability of the series deduced from the Euler-MacLaurin formula. In the second section we use finite differences and Newton series to give a convergent version of the Ramanujan summation which generalizes the classical Laplace-Gregory formula. In the third section we use the Euler-Boole summation formula to link the Ramanujan summation of even and odd terms of a series with the Euler summation of the corresponding alternate series.

4.1 A Borel Summable Series

4.1.1 A Formal Transform

In the preceding chapters we have seen that the Ramanujan summation of a series $\sum_{n\geq1} f(n)$ is related to the series involving the Bernoulli numbers

$$\sum_{k\geq1} \frac{B_k}{k!} \partial^{k-1} f(1)$$

by the formulas

$$\sum_{n\geq1}^{\mathcal{R}} f(n) = -\sum_{k=1}^{m} \frac{B_k}{k!} \partial^{k-1} f(1) + \int_1^{+\infty} \frac{b_{m+1}(x)}{(m+1)!} \partial^{m+1} f(x) dx$$

© Springer International Publishing AG 2017
B. Candelpergher, *Ramanujan Summation of Divergent Series*,
Lecture Notes in Mathematics 2185, DOI 10.1007/978-3-319-63630-6_4

or

$$\overset{\mathcal{R}}{\underset{n\geq 1}{\sum}} f(n) = -\sum_{k=1}^{m} \frac{B_k}{k!} \partial^{k-1} f(1) + (1)^{m+1} \int_0^1 R_{\partial^m f}(t+1) \frac{B_m(t)}{m!} dt$$

The series $\sum_{k\geq 1} \frac{B_k}{k!} \partial^{k-1} f(1)$ appears more directly in an operator setting of the difference equation. Let E be the *shift operator* on functions defined by

$$Eg(x) = g(x+1)$$

by the Taylor formula we get formally

$$E = e^{\partial}$$

and we can write the difference equation $R_f(x) - R_f(x+1) = f(x)$ in the form

$$(I - e^{\partial})R_f = f$$

which gives formally

$$R_f = \frac{I}{I-E} f = -\frac{\partial}{e^{\partial}-1} \partial^{-1} f = -\partial^{-1} f - \sum_{k\geq 1} \frac{B_k}{k!} \partial^{k-1} f$$

If we interpret the operator ∂^{-1} by

$$\partial^{-1} f(x) = \int_1^x f(t) dt$$

we get formally

$$R_f(x) = -\int_1^x f(t) dt - \sum_{k\geq 0} \partial^k f(x) \frac{B_{k+1}}{(k+1)!}$$

$$= -\int_1^x f(t) dt + f(x) - \sum_{k\geq 0} \partial^k f(x) \frac{(-1)^{k+1} B_{k+1}}{(k+1)!}$$

this last series is noted $\mathfrak{S}(x)$ (Hardy 1949, p.341). Taking $x = 1$ we get formally

$$\overset{\mathcal{R}}{\underset{n\geq 1}{\sum}} f(n) = f(1) - \mathfrak{S}(1) = f(1) - \sum_{k\geq 0} \partial^k f(1) \frac{(-1)^{k+1} B_{k+1}}{(k+1)!}$$

This last series is often divergent, but we can give a meaning to this sum by the Borel summation procedure.

4.1.2 Borel Summation

Let $\sum_{n\geq 0} a_n$ be a series of complex numbers and $S_n = \sum_{k=0}^{n-1} a_k$ the partial sums (we set $S_0 = 0$). As in the Abel summation method we can try to define a summation of this series by taking a generalized limit of the partial sums.

We suppose that the series $\sum_{n\geq 0} \frac{t^n}{n!} S_n$ is convergent for every $t > 0$ and we set

$$\sum_{n\geq 0}^{B} a_n = \lim_{t \to +\infty} e^{-t} \sum_{n=0}^{+\infty} \frac{t^n}{n!} S_n \text{ when this limit is finite.}$$

Since the series $\sum_{n\geq 0} \frac{t^n}{n!} S_n$ is supposed to be convergent for every $t > 0$, the function

$$s : t \mapsto e^{-t} \sum_{n=0}^{+\infty} \frac{t^n}{n!} S_n$$

is an entire function and we have

$$s'(t) = -e^{-t} \sum_{n=0}^{+\infty} \frac{t^n}{n!} S_n + e^{-t} \sum_{n=1}^{+\infty} \frac{t^{n-1}}{(n-1)!} S_n$$

$$= e^{-t} \sum_{n=0}^{+\infty} \frac{t^n}{n!} (S_{n+1} - S_n)$$

$$= e^{-t} \sum_{n=0}^{+\infty} \frac{t^n}{n!} a_n$$

Since $s(0) = S_0 = 0$, we get

$$s(t) = \int_0^t e^{-x} \Big(\sum_{n=0}^{+\infty} \frac{x^n}{n!} a_n \Big) dx$$

thus we have

$$\sum_{n=1}^{B} a_n = \int_0^{+\infty} e^{-x} \Big(\sum_{n=0}^{+\infty} \frac{x^n}{n!} a_n \Big) dx \text{ when this integral is convergent.}$$

This definition is a little too restrictive since it supposes the convergence of the series $\sum_{n\geq 0} \frac{x^n}{n!} a_n$ for every x. We can give a more general definition of the Borel summation if we use analytic continuation:

We say that the series $\sum_{n\geq 0} a_n$ is *Borel summable* if the power series $\sum_{n\geq 0} \frac{x^n}{n!} a_n$ has a radius of convergence $R > 0$ and defines by analytic continuation an analytic function $x \mapsto a(x)$ in a neighbourhood of $[0, +\infty[$ such that the integral $\int_0^{+\infty} e^{-x} a(x) dx$ is convergent.

In this case we set

$$\sum_{n\geq 0}^{\mathcal{B}} a_n = \int_0^{+\infty} e^{-x} a(x) dx$$

Example Consider the series $\sum_{n\geq 0} a_n = \sum_{n\geq 0} (-1)^n n!$, then for $|x| < 1$ we have

$$\sum_{n=0}^{+\infty} a_n \frac{x^n}{n!} = \sum_{n=0}^{+\infty} (-1)^n x^n = \frac{1}{1+x}$$

By analytic continuation we have $a(x) = \frac{1}{1+x}$ and we set

$$\sum_{n\geq 0}^{\mathcal{B}} (-1)^n n! = \int_0^{+\infty} e^{-x} \frac{1}{1+x} dx$$

Similarly for $Re(z) > 0$ we get

$$\sum_{n\geq 0}^{\mathcal{B}} \frac{(-1)^n n!}{z^{n+1}} = \int_0^{+\infty} e^{-x} \frac{1}{z+x} dx = \int_0^{+\infty} e^{-xz} \frac{1}{1+x} dx$$

Remark As in the preceding example we see that with some hypothesis the Borel summation of the series $\sum_{n\geq 0} \frac{a_n}{z^{n+1}}$ is related to the Laplace transform of function $x \mapsto a(x)$ by

$$\int_0^{+\infty} e^{-xz} a(x) dx = \sum_{n\geq 0}^{\mathcal{B}} \frac{a_n}{z^{n+1}}$$

For a function f of moderate growth this gives a Borel-sum expression of $\sum_{n\geq 1}^{\mathcal{R}} f(n) e^{-nz}$ for $0 < z < \pi$.

Since we have

$$\sum_{n\geq 1}^{\mathcal{R}} f(n) e^{-nz} = \sum_{n=1}^{+\infty} f(n) e^{-nz} - e^{-z} \int_0^{+\infty} f(x+1) e^{-xz} dx$$

and by the Taylor expansion

$$f(x+1) = \sum_{n=0}^{+\infty} \partial^n f(1) \frac{x^n}{n!}$$

we get

$$\sum_{n \geq 1}^{\mathcal{R}} f(n) e^{-nz} = \sum_{n=1}^{+\infty} f(n) e^{-nz} - e^{-z} \sum_{n \geq 0}^{\mathcal{B}} \frac{\partial^n f(1)}{z^{n+1}}$$

Taking for example $f(x) = \psi(x+1) + \gamma$ we get for $0 < z < \pi$

$$\sum_{n \geq 1}^{\mathcal{R}} H_n e^{-nz} = -\frac{Log(1 - e^{-z})}{1 - e^{-z}} - \frac{e^{-z}}{z} - e^{-z} \sum_{n \geq 1}^{\mathcal{B}} \frac{(-1)^n n!}{z^{n+1}} (\zeta(n+1) - 1)$$

4.1.3   Borel Summability of Euler-MacLaurin Series

Theorem 16 *Let a function f be analytic for $Re(z) > a$, $0 < a < 1$, such that there is $C > 0$ and $s > 0$ with*

$$|f(z)| \leq C|z|^{-s}$$

Then the series

$$\mathfrak{S}(1) = \sum_{k \geq 0} \partial^k f(1) \frac{(-1)^{k+1} B_{k+1}}{(k+1)!}$$

is Borel summable and we have

$$\sum_{n \geq 1}^{\mathcal{R}} f(n) = f(1) - \sum_{k \geq 0}^{\mathcal{B}} \partial^k f(1) \frac{(-1)^{k+1} B_{k+1}}{(k+1)!} \qquad (4.1)$$

Proof Since the proof of this theorem is given in the book "Divergent Series" of Hardy we just give a sketch of the proof.

For the Borel summability we must consider the series

$$a(t) = \sum_{k=0}^{+\infty} \partial^k f(1) \frac{(-1)^{k+1} B_{k+1}}{(k+1)!} \frac{t^k}{k!}$$

and prove that it defines, by analytic continuation, a function such that the integral $\int_0^{+\infty} e^{-t} a(t) dt$ is convergent. To get a simple expression for this function we use an integral expression of the derivatives $\partial^k f(1)$.

Let $a < \delta < 1$, by the Cauchy theorem we write

$$\frac{\partial^k f(1)}{k!} = \frac{1}{2i\pi} \int_{\delta - i\infty}^{\delta + i\infty} \frac{f(u)}{(u-1)^{k+1}}$$

Then we get

$$a(t) = \frac{1}{2i\pi} \int_{\delta-i\infty}^{\delta+i\infty} f(u) \Big(\frac{1/(1-u)}{e^{t/(1-u)} - 1} - \frac{1}{t} \Big) du$$

This gives the analytic continuation of the function $t \mapsto a(t)$ and we have

$$\int_0^{+\infty} e^{-t} a(t) dt = \frac{1}{2i\pi} \int_{\delta-i\infty}^{\delta+i\infty} f(u) \Big(\int_0^{+\infty} e^{-t} \Big(\frac{1/(1-u)}{e^{t/(1-u)} - 1} - \frac{1}{t} \Big) dt \Big) du$$

Now we use the formula

$$\int_0^{+\infty} e^{-t} \Big(\frac{\omega}{e^{\omega t} - 1} - \frac{1}{t} \Big) dt = -Log(\omega) - \psi\Big(1 + \frac{1}{\omega}\Big)$$

and we get the Borel summability with

$$\sum_{k\geq 0}^{\mathcal{B}} \partial^k f(1) \frac{(-1)^{k+1} B_{k+1}}{(k+1)!} = \frac{1}{2i\pi} \int_{\delta-i\infty}^{\delta+i\infty} f(u) \big(Log(1-u) - \psi(2-u) \big) du$$

It remains to prove (4.1), that is

$$\sum_{n\geq 1}^{\mathcal{R}} f(n) = f(1) - \frac{1}{2i\pi} \int_{\delta-i\infty}^{\delta+i\infty} f(u) \big(Log(1-u) - \psi(2-u) \big) du$$

or equivalently

$$\sum_{n\geq 1}^{\mathcal{R}} f(n) = f(1) + \frac{1}{2i\pi} \int_{\delta-i\infty}^{\delta+i\infty} f(u) \Big(\psi(1-u) - Log(1-u) - \frac{1}{u-1} \Big) du$$

$$= \frac{1}{2i\pi} \int_{\delta-i\infty}^{\delta+i\infty} f(u) \big(\psi(1-u) - Log(1-u) \big) du$$

Since by (1.34)

$$\sum_{n\geq 1}^{\mathcal{R}} \frac{1}{u-n} = \Psi(1-u) - Log(1-u)$$

we have to prove that

$$\sum_{n\geq 1}^{\mathcal{R}} f(n) = \frac{1}{2i\pi} \int_{\delta-i\infty}^{\delta+i\infty} f(u) \Big(\sum_{n\geq 1}^{\mathcal{R}} \frac{1}{u-n} \Big) du$$

and by the Cauchy theorem it is equivalent to prove that

$$\sum_{n\geq 1}^{\mathcal{R}} \frac{1}{2i\pi} \int_{\delta-i\infty}^{\delta+i\infty} f(u)\frac{1}{u-n}du = \frac{1}{2i\pi} \int_{\delta-i\infty}^{\delta+i\infty} f(u)(\sum_{n\geq 1}^{\mathcal{R}} \frac{1}{u-n})du$$

which is an immediate consequence of Theorem 12 applied to the function $g(x, u) = f(u)\frac{1}{u-x}$. $\square$

Remarks

(1) As it is observed in Hardy's book "Divergent series", the result of the preceding theorem remains valid if the function f is such that $|f(z)| \leq |z|^c$ for some $c \in \mathbb{R}$.
(2) The preceding theorem is not valid if the function f is only analytic for $Re(z) > 1$ and in a neighbourhood of 1.

Take for example

$$f(z) = \frac{1}{1+(z-1)^2} = \sum_{k\geq 0}(-1)^k(x-1)^{2k}$$

Then in the disq $D(0, 1)$ we have $f(z) = \sum_{k=0}^{+\infty}(-1)^k(x-1)^{2k}$ and the series $\mathfrak{S}(1)$ is reduced to

$$-B_1 + 0 + 0 + 0 + \dots$$

thus *it is Borel summable* and

$$f(1) - \sum_{k\geq 0}^{\mathcal{B}} \partial^k f(1)\frac{(-1)^{k+1}B_{k+1}}{(k+1)!} = \frac{1}{2}$$

But the Borel summability of the series $\mathfrak{S}(1)$ does not necessarily imply the validity of (4.1) since we have by (1.34)

$$\sum_{n=1}^{R} \frac{1}{1+(n-1)^2} = \frac{1}{2i}(\sum_{n=1}^{R} \frac{1}{n-(1+i)} - \sum_{n=1}^{R} \frac{1}{n-(1-i)})$$

$$= \frac{1}{2i}(\psi(i) - \psi(-i)) - \frac{\pi}{2}$$

$$= 0.5058777206$$

Example For $f(z) = \frac{1}{z^s}$ we get for $s \neq 1$

$$\zeta(s) = \frac{1}{s-1} + \frac{1}{2} + \sum_{k\geq 1}^{\mathcal{B}} \frac{B_{k+1}}{(k+1)!}s(s+1)\dots(s+k-1)$$

and for $s = 1$ we have

$$\gamma = \frac{1}{2} + \sum_{k \geq 1}^{\mathcal{B}} \frac{B_{k+1}}{k+1}$$

4.2 A Convergent Expansion

4.2.1 Bernoulli Numbers of Second Kind

The Borel summation formula (4.1) is not very practical for numerical evaluation since the Borel summation of a series $\sum_{k \geq 0} a_k$ involves analytic continuation of the function $x \mapsto a(x)$ given, near $x = 0$, by $a(x) = \sum_{k=0}^{+\infty} a_k \frac{x^k}{k!}$.

To obtain a more useful formula involving convergent series we use in place of the operator ∂ the *difference operator* Δ defined on functions by

$$\Delta g(x) = g(x+1) - g(x)$$

The difference equation $R_f(x) - R_f(x+1) = f(x)$ is translated in the form

$$-\Delta R_f = f$$

which gives formally

$$R_f = -\frac{I}{\Delta} f$$

Now we need an expansion of $\frac{I}{\Delta}$ similar to the expansion of $\frac{\partial}{e^\partial - I}$ in term of Bernoulli numbers. To get such an expansion we use the fact that $\Delta = e^\partial - I$ thus

$$\frac{I}{\log(I + \Delta)} = \frac{I}{\log(e^\partial)} = \partial^{-1}$$

and

$$-\frac{I}{\Delta} = -\partial^{-1} + \frac{I}{\log(I + \Delta)} - \frac{I}{\Delta}$$

we are thus led to the following definition.

Definition We define the *Bernoulli numbers of second kind* β_n by

$$\frac{t}{\log(1+t)} = \sum_{n \geq 0} \frac{\beta_n}{n!} t^n = 1 + \sum_{n \geq 0} \frac{\beta_{n+1}}{(n+1)!} t^{n+1}$$

We get

$$R = -\frac{I}{\Delta}f = -\partial^{-1}f + \sum_{n\geq 0}\frac{\beta_{n+1}}{(n+1)!}\Delta^n f$$

This gives formally

$$\sum_{n\geq 1}^{\mathcal{R}}f(n) = \sum_{n\geq 0}\frac{\beta_{n+1}}{(n+1)!}(\Delta^n f)(1) \tag{4.2}$$

In the following section we show that this series is often convergent and gives a numerical evaluation for the Ramanujan summation.

Remark The Bernoulli numbers of second kind β_n are given by $\beta_0 = 1$ and the relation

$$\sum_{k=0}^{n}\frac{\beta_k}{k!}\frac{(-1)^k}{n-k+1} = 0$$

this gives

$$\beta_1 = \frac{1}{2}, \ \beta_2 = -\frac{1}{6}, \ \beta_3 = \frac{1}{4}, \ \beta_4 = -\frac{19}{30}, \ \beta_5 = \frac{9}{4}, \ \beta_6 = -\frac{863}{84}, \dots$$

We can give a simple integral expression of the Bernoulli numbers of second kind. It suffices to write

$$\frac{t}{\log(1+t)} = \int_0^1(1+t)^x dx = \sum_{n\geq 0}t^n\int_0^1\frac{x(x-1)\dots(x-n+1)}{n!}dx$$

thus we get for $n \geq 1$

$$\beta_n = \int_0^1 x(x-1)\dots(x-n+1)dx \tag{4.3}$$

4.2.2 Newton Interpolation Formula

By the relations (4.2) and (4.3) we see that the Ramanujan summation can be related to the Newton interpolation series of the function f. We now turn to the basic definitions and properties of these series.

The Newton interpolation series are series of type

$$\sum_{n\geq 0} a_n \frac{(z-1)\dots(z-n)}{n!}$$

Where we always use the convention: $(z-1)\dots(z-n) = 1$ if $n = 0$.

These series have a "half plane convergence" property given by the following theorem (Nörlund 1926).

Theorem 1 (of Nörlund) *Let* $x_0 < 1$. *If the series* $\sum_{n\geq 0} a_n \frac{(x_0-1)\dots(x_0-n)}{n!}$ *is convergent then the series* $\sum_{n\geq 0} a_n \frac{(z-1)\dots(z-n)}{n!}$ *is uniformly convergent on every compact of the half plane* $\{Re(z) > x_0\}$. *The function*

$$f(z) = \sum_{n=0}^{+\infty} a_n \frac{(z-1)\dots(z-n)}{n!}$$

is analytic for $Re(z) > x_0$ *and there exists* $C > 0$ *such that*

$$|f(z)| \leq Ce^{|z|\frac{\pi}{2}} |z|^{x_0+\frac{1}{2}}$$

With the hypothesis of the preceding theorem we see that the coefficients a_n are related to the values $f(1), f(2), \dots f(k), \dots$ of the function f by

$$f(k+1) = \sum_{n=0}^{k} a_n \frac{k(k-1)\dots(k-n+1)}{n!} = \sum_{n=0}^{k} a_n C_k^n$$

We can invert this relation to get an expression of the coefficients a_n in terms of the values $f(1), f(2), \dots$ of the function f, we get

$$a_n = \sum_{k=0}^{n} f(k+1) C_n^k (-1)^{n-k}$$

This last formula is related to the $\Delta = E - I$ operator by

$$\Delta^n f(1) = (E-I)^n(f)(1) = \sum_{k=0}^{n} E^k(f)(1) C_n^k (-1)^{n-k} = \sum_{k=0}^{n} f(k+1) C_n^k (-1)^{n-k}$$

Thus we have

$$a_n = \Delta^n f(1)$$

This result gives the solution of the interpolation problem of finding an analytic function f given the values $f(1), f(2), \dots f(k), \dots$.

If these values satisfy the condition that the series

$$\sum_{n \geq 0} \Delta^n f(1) \frac{(x_0 - 1) \ldots (x_0 - n)}{n!} \text{ is convergent for some } x_0 < 1$$

then the analytic function f is defined for $z \in \{Re(z) > x_0\}$ by

$$f(z) = \sum_{n=0}^{+\infty} \Delta^n f(1) \frac{(z - 1) \ldots (z - n)}{n!} \tag{4.4}$$

which is the *Newton interpolation formula*.

The following theorem (Nörlund 1926) gives a simple growth condition on a function f to obtain the validity of the Newton interpolation formula.

Theorem 2 (of Nörlund) *Let $x_0 < 1$. If a function f is analytic for $Re(z) > x_0$ and verifies*

$$|f(z)| \leq C e^{|z| Log(2)}$$

for a constant $C > 0$, then for $Re(z) > \sup(x_0, 1/2)$ we have

$$f(z) = \sum_{n=0}^{+\infty} \Delta^n f(1) \frac{(z - 1) \ldots (z - n)}{n!}$$

this series is uniformly convergent for $Re(z) > \sup(x_0, 1/2) + \varepsilon \ (\varepsilon > 0)$.

4.2.3 Evaluation of $\Delta^n f(1)$

To get the expansion of a function f in Newton series we have to evaluate the terms $\Delta^n f(1)$, this can be done by

$$\Delta^n f(1) = \sum_{k=0}^{n} f(k + 1) C_n^k (-1)^{n-k} \tag{4.5}$$

but it is often more simple to use formal power series and find an explicit expression of the *exponential generating series*

$$\sum_{n \geq 0} \Delta^n f(1) \frac{t^n}{n!}.$$

We have a simple expression of this series, since

$$\sum_{n\geq 0} \Delta^n f(1)\frac{t^n}{n!} = \sum_{n\geq 0}\sum_{k=0}^{n} f(k+1)\frac{(-1)^{n-k}t^n}{k!(n-k)!} = \sum_{k\geq 0} f(k+1)\frac{t^k}{k!}\sum_{l\geq 0}\frac{(-1)^l t^l}{l!}$$

thus we get

$$\sum_{n\geq 0}\Delta^n f(1)\frac{t^n}{n!} = e^{-t}\sum_{k\geq 0} f(k+1)\frac{t^k}{k!} \qquad (4.6)$$

We can also remark that by the Laplace transform of (4.6) we also obtain a simple expression of the *ordinary generating series*

$$\sum_{n\geq 0}\Delta^n f(1)z^n$$

since we have formally

$$\sum_{n\geq 0}\Delta^n f(1)z^n = \frac{1}{z}\int_0^{+\infty} e^{-\frac{t}{z}}\sum_{n\geq 0}\Delta^n f(1)\frac{t^n}{n!}dt$$

thus we get

$$\sum_{n\geq 0}\Delta^n f(1)z^n = \frac{1}{z}\int_0^{+\infty} e^{-t\frac{z+1}{z}}\sum_{k\geq 0} f(k+1)\frac{t^k}{k!}dt$$

this gives finally

$$\sum_{n\geq 0}\Delta^n f(1)z^n = \frac{1}{z+1}\sum_{k\geq 0} f(k+1)(\frac{z}{z+1})^k \qquad (4.7)$$

As a consequence of this relation we have the *reciprocity property*

$$\Delta^n f(1) = (-1)^n g(n+1) \Leftrightarrow \Delta^n g(1) = (-1)^n f(n+1) \qquad (4.8)$$

To prove this relation we observe that

$$\frac{z}{z+1} = -Z \Leftrightarrow \frac{Z}{Z+1} = -z$$

and

$$\sum_{n\geq0}(-1)^n g(n+1)z^n = \frac{1}{z+1}\sum_{k\geq0}f(k+1)(\frac{z}{z+1})^k$$

is equivalent to

$$\frac{1}{Z+1}\sum_{n\geq0}g(n+1)(\frac{Z}{Z+1})^n = \sum_{k\geq0}f(k+1)(-1)^k Z^k$$

Remark (Ramanujan's Interpolation Formula) The formula (4.6) connects the Newton interpolation formula to an interpolation formula that Ramanujan uses. If we write

$$(z-1)\dots(z-k) = (-1)^k\frac{1}{\Gamma(-z+1)}\int_0^{+\infty}e^{-t}t^{k-z}dt$$

then we have formally

$$\sum_{k\geq0}\frac{(\Delta^k f)(1)}{k!}(z-1)\dots(z-k) = \frac{1}{\Gamma(-z+1)}\int_0^{+\infty}t^{-z}e^{-t}\sum_{k\geq0}\frac{(-1)^k(\Delta^k f)(1)}{k!}t^k dt$$

Thus by (4.6) the Newton interpolation formula becomes the *Ramanujan interpolation formula* (see Appendix)

$$f(z) = \frac{1}{\Gamma(-z+1)}\int_0^{+\infty}t^{-z}\sum_{k\geq0}\frac{(-1)^k}{k!}f(k+1)t^k dt$$

Examples

(1) If $f(x) = \frac{1}{x}$ then

$$\sum_{n\geq0}\Delta^n f(1)\frac{t^n}{n!} = e^{-t}\sum_{k\geq0}\frac{t^k}{(k+1)!} = \frac{e^{-t}}{t}(e^t-1) = \frac{1-e^{-t}}{t}$$

thus

$$\Delta^n\frac{1}{x}(1) = \frac{(-1)^n}{n+1} \tag{4.9}$$

(2) It is easy to get a simple formula relating $\Delta^n\varphi_f(1)$ to $\Delta^n f(1)$. By definition of φ_f we have

$$\Delta\varphi_f(x) = \varphi_f(x+1) - \varphi_f(x) = f(x+1)$$

thus

$$\Delta\varphi_f = Ef = (E - I + I)f = \Delta f + f$$

This gives $\Delta^0 \varphi_f(1) = f(1)$ and for $n \geq 1$

$$\Delta^n \varphi_f(1) = \Delta^n f(1) + \Delta^{n-1} f(1) \qquad (4.10)$$

If we take $f(x) = \frac{1}{x}$, then we get by (4.9)

$$\Delta^n \varphi_{\frac{1}{x}}(1) = \frac{(-1)^{n+1}}{n(n+1)} \text{ for } n \geq 1 \qquad (4.11)$$

(3) There is also a simple formula to evaluate $\Delta^n(\frac{1}{x}\varphi_f)(1)$. We have

$$\Delta^n(\frac{1}{x}\varphi_f)(1) = \sum_{k=0}^{n} \left(\frac{1}{k+1} \sum_{j=1}^{k+1} f(j) \right) C_n^k (-1)^{n-k}$$

$$= \sum_{j=0}^{n} f(j+1) \left(\sum_{k=j}^{n} C_n^k (-1)^{n-k} \frac{1}{k+1} \right)$$

now it is easy to verify by induction that

$$\sum_{k=j}^{n} C_n^k (-1)^{n-k} \frac{1}{k+1} = (-1)^{n-j} C_n^j \frac{1}{n+1}$$

thus we get

$$\Delta^n(\frac{1}{x}\varphi_f)(1) = \frac{1}{n+1} \Delta^n f(1) \qquad (4.12)$$

If $f(x) = \frac{1}{x}$, then by (4.9) we get

$$\Delta^n(\frac{1}{x}\varphi_{\frac{1}{x}})(1) = \frac{1}{n+1} \Delta^n \frac{1}{x}(1) = \frac{(-1)^n}{(n+1)^2} \qquad (4.13)$$

and by the reciprocity property (4.8)

$$\Delta^n \frac{1}{x^2}(1) = (-1)^n(\frac{1}{x}\varphi_{\frac{1}{x}})(n+1) = (-1)^n \frac{H_{n+1}}{n+1} \qquad (4.14)$$

(4) Let xf be the function $x \mapsto xf(x)$ then

$$\Delta(xf) = (x+1)f(x+1) - xf(x) = x(f(x+1) - f(x)) + f(x+1)$$

thus

$$\Delta(xf) = x\Delta f + Ef$$

and by iteration we get for $n \geq 1$

$$\Delta^n(xf) = x\Delta^n f + nE\Delta^{n-1}f = (x+n)\Delta^n f + n\Delta^{n-1}f$$

This gives $\Delta^0(xf)(1) = f(1)$ and for $n \geq 1$

$$\Delta^n(xf)(1) = (n+1)\Delta^n f(1) + n\Delta^{n-1}f(1) \tag{4.15}$$

we deduce the recursion formula for an integer $k \geq 1$

$$\Delta^n(x^k f)(1) = (n+1)\Delta^n(x^{k-1}f)(1) + n\Delta^{n-1}(x^{k-1}f)(1) \tag{4.16}$$

If we take $f(x) = \varphi_{\frac{1}{x}}$ we get by (4.12)

$$\Delta^1(x\varphi_{\frac{1}{x}})(1) = 2$$

and

$$\Delta^n(x\varphi_{\frac{1}{x}})(1) = \frac{(-1)^n}{n(n-1)} \quad \text{for } n \geq 2 \tag{4.17}$$

For $f(x) = x^2\varphi_{\frac{1}{x}}$ we have

$$\Delta^1(x^2\varphi_{\frac{1}{x}})(1) = 5, \quad \Delta^2(x^2\varphi_{\frac{1}{x}})(1) = \frac{11}{2}$$

and

$$\Delta^n(x^2\varphi_{\frac{1}{x}})(1) = \frac{(-1)^{n-1}(n+2)}{n(n-1)(n-2)} \quad \text{for } n \geq 3 \tag{4.18}$$

(5) Let $z \neq 0, -1, -2, \ldots$, then we have

$$\frac{(-1)^n n!}{z(z+1)\ldots(z+n)} = \int_0^1 x^{z-1}(x-1)^n dx = \sum_{k=0}^{n} C_n^k (-1)^{n-k} \frac{1}{z+k} \tag{4.19}$$

thus

$$\Delta^n(\frac{1}{z-1+x})(1) = (-1)^n \frac{n!}{z(z+1)\dots(z+n)} = (-1)^n \frac{\Gamma(z)\Gamma(n+1)}{\Gamma(z+n+1)} \tag{4.20}$$

If $f(x) = \frac{\Gamma(z)\Gamma(x)}{\Gamma(z+x)}$ then by the reciprocity formula (4.8) we get

$$\Delta^n f(1) = (-1)^n \frac{1}{z+n} \tag{4.21}$$

4.2.4 The Convergent Transformation Formula

Let f be given by the Newton interpolation formula, then for any integer $n \geq 1$ we have

$$f(n) = \sum_{k=0}^{+\infty} \frac{(\Delta^k f)(1)}{k!}(n-1)\dots(n-k)$$

(remember that by definition we have $(n-1)\dots(n-k) = 1$ if $k = 0$)
    To evaluate the sum $\sum_{n\geq 1}^{\mathcal{R}} f(n)$ we try to prove that

$$\sum_{n\geq 1}^{\mathcal{R}} f(n) = \sum_{k=0}^{+\infty} \frac{(\Delta^k f)(1)}{k!} \sum_{n\geq 1}^{\mathcal{R}}(n-1)\dots(n-k)$$

and use the following lemma:

Lemma 17 *We have for any integer $k \geq 0$*

$$R_{(x-1)\dots(x-k)} = -\frac{1}{k+1}(x-1)\dots(x-(k+1)) + \frac{\beta_{k+1}}{k+1}$$

(remember that by definition we have $(x-1)\dots(x-k) = 1$ if $k = 0$).
    Thus

$$\sum_{n\geq 1}^{\mathcal{R}}(n-1)\dots(n-k) = \frac{\beta_{k+1}}{k+1} \tag{4.22}$$

Proof Note that

$$(x-1)\dots(x-(k+1)) - x(x-1)\dots(x-k) = -(k+1)(x-1)\dots(x-k)$$

thus

$$R_{(x-1)...(x-k)} = -\frac{1}{k+1}(x-1)...(x-(k+1)) + \frac{1}{k+1}\int_1^2 (x-1)...(x-(k+1))dx$$

and we get

$$\sum_{n\geq1}^{\mathcal{R}}(n-1)...(n-k) = \frac{1}{k+1}\int_1^2 (x-1)...(x-(k+1))dx$$

this gives

$$\frac{1}{k+1}\int_1^2 (x-1)...(x-(k+1))dx = \frac{1}{k+1}\int_0^1 x(x-1)...(x-k)dx = \frac{\beta_{k+1}}{k+1}$$

$\square$

Theorem 18 *Let f be analytic for Re(z) > x_0 with x_0 < 1, such that there exists C > 0 with*

$$|f(z)| \leq Ce^{|z|Log(2)}$$

If we set for k ≥ 0

$$\beta_{k+1} = \int_0^1 x(x-1)...(x-k)dx$$

and

$$\Delta^k f(1) = \sum_{j=0}^k f(j+1)C_k^j(-1)^{k-j}$$

then the series $\sum_{k\geq0} \frac{\beta_{k+1}}{(k+1)!}(\Delta^k f)(1)$ is convergent and

$$\sum_{n\geq1}^{\mathcal{R}}f(n) = \sum_{k=0}^{+\infty} \frac{\beta_{k+1}}{(k+1)!}(\Delta^k f)(1)$$

Proof By (4.22) it suffices to prove that

$$R_f(x) = \sum_{k=0}^{+\infty} \frac{(\Delta^k f)(1)}{k!}R_{(x-1)...(x-k)} \tag{4.23}$$

By Theorem 2 of Nörlund we have the following expansion

$$f(x) = \sum_{k \geq 0} \frac{(\Delta^k f)(1)}{k!}(x-1)\dots(x-k)$$

where this expansion is uniformly convergent in every compact in the half-plane $Re(z) > \alpha_0 = \sup(x_0, 1/2)$. Now to prove (4.23) we use the expression of $R_{(x-1)\dots(x-k)}$ given in the preceding lemma and we consider the series

$$\sum_{k \geq 0} \frac{(\Delta^k f)(1)}{(k+1)!}(x-1)\dots(x-(k+1))$$

If we write the general term in the form

$$\frac{(\Delta^k f)(1)}{(k+1)!}(x-1)\dots(x-(k+1)) = \frac{(\Delta^k f)(1)}{k!}(x-1)\dots(x-k)\Big[\frac{x-(k+1)}{k+1}\Big]$$

and apply the classical summation by parts then we see that this series is also convergent for $Re(x) > \alpha_0$. Then we can define for $Re(x) > \alpha_0$ an analytic function

$$R(x) = -\sum_{k=0}^{+\infty} \frac{(\Delta^k f)(1)}{(k+1)!}(x-1)\dots(x-(k+1))$$

This function satisfies the difference equation

$$R(x) - R(x+1) = f(x)$$

and by Theorem 1 of Nörlund we have

$$|R(x)| \leq Ce^{|x|\frac{\pi}{2}}|x|^{\alpha_0+\frac{1}{2}}$$

Thus the function R is in $\mathcal{O}^\pi$, and the function R_f is given by

$$R_f(x) = R(x) - \int_1^2 R(t)dt$$

Note that $R(1) = 0$ and by uniform convergence of the series defining R on the interval $[1, 2]$, we get

$$-\int_1^2 R(x)dx = \sum_{k=0}^{+\infty} \frac{(\Delta^k f)(1)}{(k+1)!}\int_1^2 (x-1)\dots(x-(k+1))dx = \sum_{k=0}^{+\infty} \frac{\beta_{k+1}}{(k+1)!}(\Delta^k f)(1)$$

□

Remark Note that for any integer $m \geq 1$ and if $g(x) = f(x + m)$, then we have

$$\Delta^n g(1) = \sum_{k=0}^{n} f(k + m + 1) C_n^k (-1)^{n-k} = \Delta^n f(m + 1)$$

and by the shift property

$$\sum_{n\geq 1}^{\mathcal{R}} f(n) = \sum_{n\geq 1}^{\mathcal{R}} g(n) + \sum_{n=1}^{m} f(n) - \int_1^{m+1} f(x)dx$$

thus for any integer $m \geq 1$ we get

$$\int_1^{m+1} f(x)dx - \sum_{n=1}^{m} f(n) = \sum_{k=0}^{+\infty} \frac{\beta_{k+1}}{(k+1)!} (\Delta^k f(m+1) - \Delta^k f(1))$$

which is the classical Laplace integration formula (also called Gregory's formula) (Boole 1960).

Replacing m by $m - 1$ we write this last formula in the form

$$\sum_{n=1}^{m} f(n) - \int_1^{m} f(x)dx = \sum_{n\geq 1}^{\mathcal{R}} f(n) + f(m) - \sum_{k=0}^{+\infty} \frac{\beta_{k+1}}{(k+1)!} \Delta^k f(m)$$

which shows the analogy with the Euler-MacLaurin expansion

$$\sum_{n=1}^{m} f(n) - \int_1^{m} f(x)dx = \sum_{n\geq 1}^{\mathcal{R}} f(n) + f(m) + \sum_{k\geq 0} \frac{B_{k+1}}{(k+1)!} \partial^k f(m)$$

Examples

(1) If we take $f(x) = \frac{1}{x}$, then by (4.9) we get

$$\gamma = \sum_{n\geq 1}^{\mathcal{R}} \frac{1}{n} = \sum_{k=0}^{+\infty} \frac{(-1)^k}{k+1} \frac{\beta_{k+1}}{(k+1)!} \tag{4.24}$$

(2) For $f = \varphi_{\frac{1}{x}}$ we have $f(n) = H_n$, and we get by (4.11)

$$\sum_{n\geq 1}^{\mathcal{R}} H_n = \frac{1}{2} + \sum_{k=1}^{+\infty} \frac{(-1)^{k+1}}{k(k+1)} \frac{\beta_{k+1}}{(k+1)!} \tag{4.25}$$

By (3.3) we get

$$\frac{3}{2}\gamma - Log(\sqrt{2\pi}) = \sum_{k=1}^{+\infty} \frac{(-1)^{k+1}}{k(k+1)} \frac{\beta_{k+1}}{(k+1)!}$$

with (4.24) this gives

$$Log(\sqrt{2\pi}) = \frac{3}{4} + \frac{1}{2}\sum_{k=1}^{+\infty} \frac{(-1)^k(3k+2)}{k(k+1)} \frac{\beta_{k+1}}{(k+1)!} \tag{4.26}$$

(3) For $f = x\varphi_{\frac{1}{x}}$, we have $f(n) = nH_n$, and we get by (4.17)

$$\sum_{n\geq 1}^{\mathcal{R}} nH_n = \frac{1}{2} - \frac{1}{6} + \sum_{k=2}^{+\infty} \frac{(-1)^k}{k(k-1)} \frac{\beta_{k+1}}{(k+1)!} \tag{4.27}$$

By (3.4) we get

$$\zeta'(-1) = \frac{5}{12}\gamma - Log(\sqrt{2\pi}) + \frac{13}{24} - \sum_{k=2}^{+\infty} \frac{(-1)^k}{k(k-1)} \frac{\beta_{k+1}}{(k+1)!}$$

With (4.21) and (4.23) this gives

$$\zeta'(-1) = -\frac{25}{288} - \frac{1}{12}\sum_{k=2}^{+\infty} \frac{(-1)^k(13k+11)}{(k-1)(k+1)} \frac{\beta_{k+1}}{(k+1)!}$$

By the same type of calculations we get also

$$\sum_{n\geq 1}^{\mathcal{R}} n^2 H_n = \frac{15}{48} + \sum_{k=3}^{+\infty} \frac{(-1)^{k-1}(k+2)}{k(k-1)(k-2)} \frac{\beta_{k+1}}{(k+1)!} \tag{4.28}$$

which gives by (3.5)

$$\zeta'(-2) = -\frac{1}{3}\gamma + Log(\sqrt{2\pi}) + 2\zeta'(-1) - \frac{3}{16} + \sum_{k=3}^{+\infty} \frac{(-1)^{k-1}(k+2)}{k(k-1)(k-2)} \frac{\beta_{k+1}}{(k+1)!}$$

Similar formulas can be obtained for $\sum_{n\geq 1}^{\mathcal{R}} n^k H_n$, which gives expansions of $\zeta'(-2k)$, thus of $\zeta(2k+1)$ (Coppo and Young 2016).

(4) By (4.13) we get

$$\sum_{n\geq 1}^{\mathcal{R}} \frac{H_n}{n} = \sum_{k=0}^{+\infty} \frac{\beta_{k+1}}{(k+1)!} \frac{(-1)^k}{(k+1)^2} \tag{4.29}$$

And by (4.14)

$$\frac{\pi^2}{6} = 1 + \sum_{n\geq 1}^{\mathcal{R}} \frac{1}{n^2} = 1 + \sum_{k=0}^{+\infty} \frac{\beta_{k+1}}{(k+1)!} \frac{(-1)^k H_{k+1}}{(k+1)} \tag{4.30}$$

(5) By (4.21) we get for $z \neq 0, -1, -2, \ldots$

$$\sum_{n\geq 1}^{\mathcal{R}} \frac{\Gamma(z)\Gamma(n)}{\Gamma(z+n)} = \sum_{k=0}^{+\infty} \frac{\beta_{k+1}}{(k+1)!} (-1)^k \frac{1}{z+k} \tag{4.31}$$

Since it is immediately verified that for $\mathrm{Re}(z) > 1$ we have

$$R_{\frac{\Gamma(z)\Gamma(x)}{\Gamma(z+x)}} = \frac{\Gamma(z-1)\Gamma(x)}{\Gamma(z+x-1)} - \int_1^2 \frac{\Gamma(z-1)\Gamma(x)}{\Gamma(z+x-1)} dx$$

this gives

$$\sum_{n\geq 1}^{\mathcal{R}} \frac{\Gamma(z)\Gamma(n)}{\Gamma(z+n)} = \frac{1}{z-1} - \int_1^2 \frac{\Gamma(z-1)\Gamma(x)}{\Gamma(z+x-1)} dx \tag{4.32}$$

thus we get in this case

$$\sum_{k=0}^{+\infty} \frac{\beta_{k+1}}{(k+1)!} (-1)^k \frac{1}{z+k} = \frac{1}{z-1} - \int_1^2 \frac{\Gamma(z-1)\Gamma(x)}{\Gamma(z+x-1)} dx$$

(6) By (4.20) we get for $z \neq 0, -1, -2, \ldots$

$$\sum_{n\geq 1}^{\mathcal{R}} \frac{1}{z-1+n} = \sum_{k=0}^{+\infty} \frac{\beta_{k+1}}{(k+1)!} (-1)^k \frac{k!}{z(z+1)\ldots(z+k)}$$

which gives by (2.4)

$$\mathrm{Log}(z) - \psi(z) = \sum_{k=0}^{+\infty} \frac{\beta_{k+1}}{(k+1)!} \frac{(-1)^k k!}{z(z+1)\ldots(z+k)} \tag{4.33}$$

Remark Let $f \in \mathcal{O}^{\pi}$ and $F(x) = \int_1^x f(t)dt$, we have by the shift property

$$\sum_{n\geq 1}^{\mathcal{R}} \Delta^k f(n) = \sum_{j=0}^{k} C_j^k (-1)^{k-j} \sum_{n\geq 1}^{\mathcal{R}} f(j+n)$$

$$= \sum_{j=0}^{k} C_j^k (-1)^{k-j} \left(\sum_{n\geq 1}^{\mathcal{R}} f(n) - \varphi_f(j+1) + f(j+1) + F(j+1) \right)$$

that is

$$\sum_{n\geq 1}^{\mathcal{R}} \Delta^k f(n) = -\Delta^k \varphi_f(1) + \Delta^k f(1) + \Delta^k F(1)$$

this gives by (4.10), for $k \geq 1$, a formula that generalizes the shift property

$$\sum_{n\geq 1}^{\mathcal{R}} \Delta^k f(n) = -\Delta^{k-1} f(1) + \Delta^k F(1) \tag{4.34}$$

For $f(x) = \frac{1}{x}$ we have by (4.19)

$$\Delta^k \frac{1}{x}(z) = \frac{(-1)^k k!}{z(z+1)\ldots(z+k)}$$

thus by (4.9) we have for $k \geq 1$

$$\sum_{n\geq 1}^{\mathcal{R}} \frac{(-1)^k k!}{n(n+1)\ldots(n+k)} = (-1)^k \frac{1}{k} + \Delta^k Log(1)$$

For $k \geq 1$ the series $\sum_{n\geq 1}^{\mathcal{R}} \frac{(-1)^k k!}{n(n+1)\ldots(n+k)}$ is convergent and since

$$\frac{1}{n(n+1)\ldots(n+k-1)} - \frac{1}{(n+1)\ldots(n+k)} = \frac{k}{n(n+1)\ldots(n+k)}$$

we have

$$\sum_{n=1}^{+\infty} \frac{k!}{n(n+1)\ldots(n+k)} = \frac{1}{k}$$

Thus for $k \geq 1$ we get

$$\Delta^k Log(1) = (-1)^{k-1} \int_1^{+\infty} \frac{k!}{x(x+1)\ldots(x+k)} dx$$

4.3 Summation of Alternating Series

4.3.1 Euler Summation of Alternating Series

Consider the Euler polynomials $E_n(x)$ defined by the generating function

$$\sum_{n \geq 0} \frac{E_n(x)}{n!} z^n = \frac{2e^{xz}}{e^z + 1}$$

We define the numbers $E_n = E_n(0)$

$$E_0 = 1, E_1 = -\frac{1}{2}, E_2 = 0, E_3 = \frac{1}{4}, E_4 = 0, E_5 = -\frac{1}{2}, \ldots$$

These numbers are related to the Bernoulli numbers by

$$E_n = \frac{-2(2^{n+1} - 1)}{n+1} B_{n+1} \quad \text{for } n \geq 1$$

(we don't call these numbers "Euler numbers" since the Euler numbers are usually defined by $2^n E_n(\frac{1}{2})$).

With the generating function we verify that $E_n(1-x) = (-1)^n E_n(x)$ for $n \geq 0$, and we define for $t \in \mathbb{R}$ the functions

$$e_m(t) = (-1)^{[t]}(-1)^m E_m(t - [t])$$

We can define the summation of alternating series $\sum_{n \geq 1}(-1)^n f(n)$ starting from an analogue of the Euler-MacLaurin formula that is the Euler-Boole summation formula (cf. Appendix)

$$f(1) - f(2) + \ldots + (-1)^{n-1} f(n) = \frac{1}{2} \sum_{k=0}^{m} \partial^k f(1) \frac{E_k}{k!}$$

$$+ \frac{(-1)^{n-1}}{2} \sum_{k=0}^{m} \partial^k f(n+1) \frac{E_k}{k!}$$

$$+ \frac{1}{2} \int_1^{n+1} \frac{1}{m!} e_m(t) \, \partial^{m+1} f(t) \, dt$$

which we can write in the form

$$f(1) - \ldots + (-1)^{n-1} f(n) = \frac{1}{2} \sum_{k=0}^{m} \partial^k f(1) \frac{E_k}{k!} + \frac{1}{2} \int_{1}^{+\infty} \frac{1}{m!} e_m (t) \, \partial^{m+1} f (t) \, dt$$

$$+ \frac{(-1)^{n-1}}{2} \sum_{k=0}^{m} \partial^k f(n+1) \frac{E_k}{k!}$$

$$- \frac{1}{2} \int_{n+1}^{\infty} \frac{1}{m!} e_m (t) \, \partial^{m+1} f (t) \, dt$$

We see that this formula has the same structure as the Euler-MacLaurin formula, with the constant term

$$\tilde{C}(f) = \frac{1}{2} \sum_{k=0}^{m} \partial^k f(1) \frac{E_k}{k!} + \frac{1}{2} \int_{1}^{+\infty} \frac{1}{m!} e_m (t) \, \partial^{m+1} f (t) \, dt$$

which is, by integration by parts, independent of m for $m \geq M$.

We can proceed, for an alternating series $\sum_{k \geq 1} (-1)^{k-1} f(k)$, as in the Ramanujan summation, defining the *Euler summation* of this series by

$$\sum_{k \geq 1}^{\mathcal{E}} (-1)^{k-1} f(k) = \tilde{C}(f) = \frac{1}{2} \sum_{k=0}^{m} \partial^k f(1) \frac{E_k}{k!} + \frac{1}{2} \int_{1}^{+\infty} \frac{1}{m!} e_m (t) \, \partial^{m+1} f (t) \, dt$$

As in the Ramanujan summation it is useful to give a definition of the Euler summation not directly dependent on the Euler-Boole summation formula.

If we set

$$T_f(n) = \frac{(-1)^n}{2} \sum_{k=0}^{m} \partial^k f(n) \frac{E_k}{k!} - \frac{1}{2} \int_{n}^{\infty} \frac{1}{m!} e_m (t) \, \partial^{m+1} f (t) \, dt$$

then the Euler-Boole summation formula is

$$f(1) - \ldots + (-1)^{n-1} f(n) = \tilde{C}(f) + T_f(n+1)$$

thus

$$(-1)^{n-1} f(n) = T_f(n+1) - T_f(n)$$

Now if we define the sequence A_f by

$$A_f(n) = (-1)^n T_f(n)$$

we get

$$A_f(n) + A_f(n+1) = f(n) \tag{4.35}$$

and

$$\sum_{n \geq 1}^{\mathcal{E}} (-1)^{n-1} f(n) = \tilde{C}(f) = -T_f(1) = A_f(1) \tag{4.36}$$

By (4.23) and (4.24) it is natural to define the Euler summation by

$$\sum_{n \geq 1}^{\mathcal{E}} (-1)^{n-1} f(n) = \tilde{C}(f) = A_f(1)$$

where

$$A_f(x) + A_f(x+1) = f(x)$$

But this equation does not specify a unique function A_f since we must avoid the solutions of the equation $A(x) + A(x+1) = 0$.

Lemma 19 *If $f \in \mathcal{O}^\pi$ there exists a unique solution $A_f \in \mathcal{O}^\pi$ of*

$$A_f(x) + A_f(x+1) = f(x)$$

and we have

$$A_f(x) = R_{f(2x)}\left(\frac{x}{2}\right) - R_{f(2x)}\left(\frac{x+1}{2}\right). \tag{4.37}$$

Proof Uniqueness of the solution: if a function $A \in \mathcal{O}^\pi$ is a solution of the equation $A(x) + A(x+1) = 0$ then the function $R(x) = A(x)e^{i\pi x}$ is a solution of the equation $R(x) - R(x+1) = 0$, and is of exponential type $< 2\pi$, thus by Lemma 1 the function R is a constant C and we have $A(x) = Ce^{i\pi x}$, and $A \in \mathcal{O}^\pi$ implies $C = 0$.

Existence of the solution: since the function $x \mapsto f(2x)$ is in $\mathcal{O}^{2\pi}$, then by Theorem 1 the function $R_{f(2x)} \in \mathcal{O}^{2\pi}$ is a solution of the equation

$$R_{f(2x)}(x) - R_{f(2x)}(x+1) = f(2x) \text{ with } \int_1^2 R_{f(2x)}(x)dx = 0$$

If the function A_f is defined by (4.37) then we have

$$A_f(x) + A_f(x+1) = -R_{f(2x)}\left(\frac{x}{2} + 1\right) + R_{f(2x)}\left(\frac{x}{2}\right) = f(x).$$

□

Definition 6 If $f \in \mathcal{O}^\pi$ there exists a unique function $A_f \in \mathcal{O}^\pi$ which satisfies

$$A_f(x) + A_f(x + 1) = f(x)$$

and we define the Euler summation of the series $\sum_{n \geq 1} (-1)^{n-1} f(n)$ by

$$\sum_{n \geq 1}^{\mathcal{E}} (-1)^{n-1} f(n) = A_f(1)$$

Remark Note that if $f \in \mathcal{O}^\pi$, then the function $g : x \mapsto f(x)e^{i\pi x}$ is in $\mathcal{O}^{2\pi}$ and by Theorem 1 there is a unique function $R_g \in \mathcal{O}^{2\pi}$ which satisfies

$$R_g(x) - R_g(x + 1) = f(x)e^{i\pi x}$$

Then the function $A(x) = e^{-i\pi x} R_g(x)$ is a solution of $A(x) + A(x+1) = f(x)$, but A is not necessarily the function A_f of the preceding definition since it is not necessarily of exponential type $< \pi$.

Examples

(1) By the generating function of Euler polynomials $E_k(x)$ we verify that

$$\sum_{k \geq 0} \frac{E_k(x) + E_k(x + 1)}{k!} z^k = \frac{2(e^{xz} + e^{(x+1)z})}{e^z + 1} = 2e^{xz}$$

thus the Euler polynomials are solutions of

$$E_k(x) + E_k(x + 1) = 2x^k \text{ for } k \geq 0$$

and by the preceding theorem we get

$$A_{x^k}(x) = \frac{1}{2} E_k(x) \text{ for } k \geq 0$$

We get for example:

$$\sum_{k \geq 1}^{\mathcal{E}} (-1)^{k-1} = \frac{1}{2}, \quad \sum_{k \geq 1}^{\mathcal{E}} (-1)^{k-1} k = \frac{1}{4}$$

More generally we have

$$\sum_{n \geq 1}^{\mathcal{E}} (-1)^{n-1} n^k = \frac{E_k}{2} = \frac{(1 - 2^{k+1})}{k + 1} B_{k+1}$$

thus for any integer $k \geq 1$ we have

$$\sum_{n \geq 1}^{\mathcal{E}} (-1)^{n-1} n^{2k} = 0$$

(2) For $f(x) = \frac{1}{x}$ we have $R_f = -\psi$ and since in this case $f(2x) = \frac{1}{2}f(x)$ we get by (4.37)

$$A_f(x) = \frac{1}{2}R_f(\frac{x}{2}) - \frac{1}{2}R_f(\frac{x+1}{2})$$

$$= \frac{1}{2}(\psi(\frac{x+1}{2}) - \psi(\frac{x}{2}))$$

Thus $A_{\frac{1}{x}}$ is the classical function

$$\beta(x) = \frac{1}{2}(\psi(\frac{x+1}{2}) - \psi(\frac{x}{2})) = \sum_{n=0}^{+\infty} \frac{(-1)^n}{n+x}$$

and we have

$$\sum_{n \geq 1}^{\mathcal{E}} \frac{(-1)^{n-1}}{n} = \beta(1) = \sum_{n=1}^{+\infty} \frac{(-1)^{n-1}}{n} = Log(2)$$

(3) If $f(x) = Log(x)$ then $R_f(x) = -Log(\Gamma(x)) + Log(\sqrt{2\pi}) - 1$, thus

$$R_{Log(2x)} = R_{Log(2)} - Log(\Gamma(x)) + Log(\sqrt{2\pi}) - 1$$

and we get

$$R_{Log(2x)} = (\frac{3}{2} - x)Log(2) - Log(\Gamma(x)) + Log(\sqrt{2\pi}) - 1$$

therefore

$$A_f(x) = \frac{1}{2}Log(2) - Log(\Gamma(x/2)) + Log(\Gamma((x+1)/2))$$

Which gives $A_f(1) = \frac{1}{2}Log(2) - Log(\Gamma(1/2))$, thus

$$\sum_{n \geq 1}^{\mathcal{E}} (-1)^{n-1} Log(n) = \frac{1}{2}Log(\frac{2}{\pi}) \qquad (4.38)$$

(4) We have for $|z| < \pi$

$$e^{ixz} + e^{i(x+1)z} = e^{ixz}(1 + e^{iz})$$

thus $A_{e^{ixz}} = \frac{e^{ixz}}{1+e^{iz}}$ and

$$\sum_{n\geq 1}^{\mathcal{E}} (-1)^{n-1} e^{inz} = \frac{e^{iz}}{1 + e^{iz}}$$

We deduce that for $-\pi < t < \pi$

$$\sum_{n\geq 1}^{\mathcal{E}} (-1)^{n-1} \cos(nt) = \frac{1}{2} \quad \text{and} \quad \sum_{n\geq 1}^{\mathcal{E}} (-1)^{n-1} \sin(nt) = \frac{1}{2} \tan(\frac{t}{2})$$

Remark Classically the Euler method of summation of a series $\sum_{n\geq 1} a_n$ is defined by (Hardy 1949)

$$\sum_{n\geq 1}^{\mathcal{E}} a_n = \sum_{k=0}^{+\infty} \frac{1}{2^{k+1}} \sum_{j=0}^{k} C_k^j a_{j+1} \quad \text{(when this last series is convergent)}$$

thus if $a_n = (-1)^{n-1} f(n)$ this gives

$$\sum_{n\geq 1}^{\mathcal{E}} a_n = \sum_{k=0}^{+\infty} \frac{(-1)^k}{2^{k+1}} \sum_{j=0}^{k} C_k^j (-1)^{k-j} f(j+1) = \sum_{k=0}^{+\infty} \frac{(-1)^k}{2^{k+1}} (\Delta^k f)(1)$$

To see how this definition is connected with our preceding definition we use the Newton expansion of the function f. If f is analytic for $Re(x) > x_0, x_0 < 1$ and such that

$$|f(x)| \leq C e^{|x|Log(2)}$$

for a constant $C > 0$, then for $Re(x) > \sup(x_0, 1/2)$ we have by Theorem 2 of Nörlund

$$f(x) = \sum_{k=0}^{+\infty} \Delta^k f(1) \frac{(x-1)\ldots(x-k)}{k!}$$

If we admit that, as in (4.16) for R_f, we have

$$A_f(x) = \sum_{k=0}^{+\infty} \Delta^k f(1) \frac{A_{(x-1)\ldots(x-k)}}{k!}$$

then it suffices to evaluate $A_{(x-1)...(x-k)}$. This is done by using the immediate result

$$A_{(1+z)^{x-1}} = \frac{1}{2} \frac{(1+z)^{x-1}}{1+\frac{z}{2}} \quad \text{for } |z| < 1$$

By the binomial expansion of $(1+z)^{x-1}$ for $|z| < 1$, and by identification, we get

$$A_{(x-1)...(x-k)} = \frac{(-1)^k k!}{2^{k+1}} \sum_{j=0}^{k} \frac{(x-1)...(x-j)}{j!}(-2)^j$$

Thus

$$A_f(x) = \sum_{k=0}^{+\infty} \Delta^k f(1) \frac{(-1)^k}{2^{k+1}} \sum_{j=0}^{k} \frac{(x-1)...(x-j)}{j!}(-2)^j$$

and with $x = 1$ we obtain

$$\sum_{n\geq 1}^{\mathcal{E}} (-1)^{n-1} f(n) = \sum_{k=0}^{+\infty} \frac{(-1)^k}{2^{k+1}} (\Delta^k f)(1) \tag{4.39}$$

that is the classical Euler summation of alternating series.

4.3.2 Properties of the Summation

The properties of the Euler summation are very easily deduced from our definition.

(1) Linearity

Since the equation defining A_f is linear we have immediately the property of linearity

$$\sum_{n\geq 1}^{\mathcal{E}} (-1)^{n-1}(af(n) + bg(n)) = a \sum_{n\geq 1}^{\mathcal{E}} (-1)^{n-1} f(n) + b \sum_{n\geq 1}^{\mathcal{E}} (-1)^{n-1} g(n)$$

(2) The translation property

Let $f \in \mathcal{O}^\pi$. Since we have

$$A_f(x+1) + A_f(x+2) = f(x+1)$$

we deduce that $A_{f(x+1)}(x) = A_f(x + 1)$, thus

$$\sum_{n\geq 1}^{\mathcal{E}} (-1)^{n-1} f(n + 1) = A_f(2) = f(1) - A_f(1)$$

which gives the usual translation property

$$\sum_{n\geq 1}^{\mathcal{E}} (-1)^{n-1} f(n + 1) = f(1) - \sum_{n\geq 1}^{\mathcal{E}} (-1)^{n-1} f(n) \qquad (4.40)$$

More generally, we have for any integer $p \geq 1$

$$A_f(x + p) + A_f(x + p + 1) = f(x + p)$$

thus if we note $f(+p) : x \mapsto f(x + p)$ then $A_{f(+p)}(x) = A_f(x + p)$ and

$$\sum_{n\geq 1}^{\mathcal{R}} (-1)^{n-1} f(n + p) = A_f(p + 1) = (-1)^p A_f(1) + (-1)^{p+1} \sum_{k=1}^{p} (-1)^{k-1} f(k)$$

Since

$$(-1)^{p+1} A_f(p + 1) + A_f(1) = \sum_{k=1}^{p} (-1)^{k-1} f(k) \qquad (4.41)$$

we get the usual translation property

$$\sum_{n\geq 1}^{\mathcal{R}} (-1)^{n-1} f(n + p) = \sum_{k=1}^{p} (-1)^{k-p} f(k) + (-1)^p \sum_{n\geq 1}^{\mathcal{E}} (-1)^{n-1} f(n)$$

(3) Relation to usual summation

Let $f \in \mathcal{O}^\pi$ and suppose that the series $\sum_{n\geq 0} (-1)^n f(x + n)$ is convergent for all $Re(x) > 0$ and defines the function

$$g : x \mapsto \sum_{n=0}^{\infty} (-1)^n f(x + n)$$

We have

$$g(x) + g(x + 1) = \sum_{n=0}^{\infty} (-1)^n f(x + n) - \sum_{n=1}^{\infty} (-1)^n f(x + n) = f(x)$$

If $g \in \mathcal{O}^{\pi}$, this function is by Lemma 17, the unique solution of the equation $g(x) + g(x + 1) = f(x)$. Thus we get

$$A_f(x) = \sum_{n=0}^{\infty} (-1)^n f(x + n)$$

and

$$\sum_{n \geq 1}^{\mathcal{E}} (-1)^{n-1} f(n) = \sum_{n=1}^{\infty} (-1)^{n-1} f(n)$$

In this case we say briefly that *"we are in a case of convergence"*.

Example By the translation property we have

$$\sum_{n \geq 1}^{\mathcal{E}} (-1)^{n-1} Log(n + 1) = - \sum_{n \geq 1}^{\mathcal{E}} (-1)^{n-1} Log(n)$$

and by (4.38) we get the well known result (Sondow 2005)

$$\sum_{n=1}^{+\infty} (-1)^{n-1} \left(\frac{1}{n} - Log\left(\frac{n + 1}{n} \right) \right) = Log\left(\frac{4}{\pi} \right)$$

Remarks

(1) Since we have $\varphi_f(n + 1) = \varphi_f(n) + f(n)$, we deduce immediately from (4.40) that

$$\sum_{n \geq 1}^{\mathcal{E}} (-1)^{n-1} \varphi_f(n) = \frac{1}{2} \sum_{n \geq 1}^{\mathcal{E}} (-1)^{n-1} f(n) \tag{4.42}$$

For example we have

$$\sum_{n \geq 1}^{\mathcal{E}} (-1)^{n-1} H_n = \frac{1}{2} Log(2) \tag{4.43}$$

Since

$$\varphi_f(n) = C_f - R_f(n) + f(n) \text{ with } C_f = \sum_{n \geq 1}^{\mathcal{R}} f(n)$$

and $\sum_{n\geq 1}^{\mathcal{E}}(-1)^{n-1} = \frac{1}{2}$, we deduce from (4.42) the following relation with the Ramanujan summation

$$\sum_{n\geq 1}^{\mathcal{R}} f(n) = 2\sum_{n\geq 1}^{\mathcal{E}}(-1)^{n-1}R_f(n) - \sum_{n\geq 1}^{\mathcal{E}}(-1)^{n-1}f(n) \tag{4.44}$$

(2) Let $f \in \mathcal{O}^\pi$ and $F(x) = \int_1^x f(t)dt$. We have

$$\int_x^{x+1} A_f(x)dx + \int_{x+1}^{x+2} A_f(x)dx = \int_x^{x+1} f(x)dx$$

thus we get

$$\sum_{n\geq 1}^{\mathcal{E}}(-1)^{n-1}\int_n^{n+1} f(x)dx = \int_1^2 A_f(x)dx$$

and by the translation property this gives

$$\sum_{n\geq 1}^{\mathcal{E}}(-1)^{n-1}F(n) = -\frac{1}{2}\int_1^2 A_f(x)dx \tag{4.45}$$

4.3.3   Relation with the Ramanujan Summation

The Ramanujan summation of even and odd terms in a divergent series $\sum_{n\geq 1} f(n)$ is simply connected with the Euler summation of the alternating series $\sum_{n\geq 1}(-1)^{n-1}f(n)$.

Theorem 20   *For* $f \in \mathcal{O}^{\pi/2}$ *we have*

$$\sum_{n\geq 1}^{\mathcal{R}} f(2n-1) = \frac{1}{2}\sum_{n\geq 1}^{\mathcal{R}} f(n) + \frac{1}{2}\sum_{n\geq 1}^{\mathcal{E}}(-1)^{n-1}f(n)$$

$$\sum_{n\geq 1}^{\mathcal{R}} f(2n) = \frac{1}{2}\sum_{n\geq 1}^{\mathcal{R}} f(n) - \frac{1}{2}\sum_{n\geq 1}^{\mathcal{E}}(-1)^{n-1}f(n) + \frac{1}{2}\int_1^2 f(t)dt$$

Proof   It is equivalent to prove the following assertions

$$\sum_{n\geq 1}^{\mathcal{R}} f(2n-1) + \sum_{n\geq 1}^{\mathcal{R}} f(2n) = \sum_{n\geq 1}^{\mathcal{R}} f(n) + \frac{1}{2}\int_1^2 f(t)dt \tag{4.46}$$

and

$$\overset{\mathcal{R}}{\underset{n\geq 1}{\sum}}f(2n-1) - \overset{\mathcal{R}}{\underset{n\geq 1}{\sum}}f(2n) = \overset{\mathcal{E}}{\underset{n\geq 1}{\sum}}(-1)^{n-1}f(n) - \frac{1}{2}\int_1^2 f(t)dt \qquad (4.47)$$

We have already proved assertion (4.46) which happens to be (2.18).
To prove (4.47) we consider the function $g : x \mapsto f(2x)$. By Lemma 19 we have

$$A_f(x) = R_g(\frac{x}{2}) - R_g(\frac{x+1}{2})$$

this gives

$$\overset{\mathcal{E}}{\underset{n\geq 1}{\sum}}(-1)^{n-1}f(n) = R_g(\frac{1}{2}) - \overset{\mathcal{R}}{\underset{n\geq 1}{\sum}}g(n)$$

But $R_g(\frac{1}{2})$ is given by (2.2) with $x = -\frac{1}{2}$, and we get

$$\overset{\mathcal{E}}{\underset{n\geq 1}{\sum}}(-1)^{n-1}f(n) = \overset{\mathcal{R}}{\underset{n\geq 1}{\sum}}g(n-\frac{1}{2}) - \overset{\mathcal{R}}{\underset{n\geq 1}{\sum}}g(n) + \int_{1/2}^1 g(t)dt$$

that is

$$\overset{\mathcal{E}}{\underset{n\geq 1}{\sum}}(-1)^{n-1}f(n) = \overset{\mathcal{R}}{\underset{n\geq 1}{\sum}}f(2n-1) - \overset{\mathcal{R}}{\underset{n\geq 1}{\sum}}f(2n) + \frac{1}{2}\int_1^2 f(t)dt$$

□

Remark For $f \in \mathcal{O}^{\pi/2}$ the preceding theorem gives

$$\overset{\mathcal{E}}{\underset{n\geq 1}{\sum}}(-1)^{n-1}f(n) = \overset{\mathcal{R}}{\underset{n\geq 1}{\sum}}f(n) - 2\overset{\mathcal{R}}{\underset{n\geq 1}{\sum}}f(2n) + \int_1^2 f(t)dt$$

$$= 2\overset{\mathcal{R}}{\underset{n\geq 1}{\sum}}f(2n-1) - \overset{\mathcal{R}}{\underset{n\geq 1}{\sum}}f(n)$$

This last formula shows that if f depends on an extra parameter z or t, then the theorems of analyticity and integration of Chap. 2 remains valid.
For example, we know that for $Re(z) < 0$

$$\overset{\mathcal{E}}{\underset{n\geq 1}{\sum}}(-1)^{n-1}e^{zn}H_n = \overset{+\infty}{\underset{n\geq 1}{\sum}}(-1)^{n-1}e^{zn}H_n = \frac{\ln(e^z+1)}{e^z+1}$$

Since this function of z is analytic near 0, by derivation with respect to z, we obtain

$$\sum_{n\geq 1}^{\varepsilon}(-1)^{n-1}nH_n = \frac{1}{4} - \frac{\ln(2)}{4}$$

$$\sum_{n\geq 1}^{\varepsilon}(-1)^{n-1}n^2H_n = -\frac{1}{16}$$

As another example, we have for $Re(s) > 1$

$$\sum_{n\geq 1}^{\varepsilon}(-1)^{n-1}\frac{1}{n^s} = \sum_{n\geq 1}^{+\infty}(-1)^{n-1}\frac{1}{n^s} = \zeta(s)(1 - 2^{1-s})$$

This function of s is analytic for $s \neq 1$, by derivation with respect to s, we have

$$\sum_{n\geq 1}^{\varepsilon}\frac{(-1)^{n-1}Log(n)}{n^s} = \zeta'(s)(2^{1-s} - 1) - \zeta(s)2^{1-s}Log(2)$$

This gives for any integer $k \geq 1$ the relation

$$\sum_{n\geq 1}^{\varepsilon}(-1)^{n-1}n^kLog(n) = 2^{k+1}\frac{B_{k+1}}{k+1}Log(2) + (2^{k+1} - 1)\zeta'(-k)$$

for $k = 1, 2$ we have

$$\sum_{n\geq 1}^{\varepsilon}(-1)^{n-1}nLog(n) = \frac{1}{3}Log(2) + 3\zeta'(-1)$$

$$\sum_{n\geq 1}^{\varepsilon}(-1)^{n-1}n^2Log(n) = 7\zeta'(-2) = \frac{-7}{4\pi^2}\zeta(3)$$

From the translation property we deduce the values of the convergent sums

$$\sum_{n=1}^{+\infty}(-1)^{n-1}(nLog(\frac{n+1}{n}) - 1) = -\frac{1}{2} - \frac{1}{6}Log(2) - \frac{1}{2}Log(\pi) - 6\zeta'(-1)$$

$$\sum_{n=1}^{+\infty}(-1)^{n-1}(n^2Log(\frac{n+1}{n}) - n + \frac{1}{2}) = \frac{Log(2)}{6} + \frac{Log(\pi)}{2} + 6\zeta'(-1) + \frac{7}{2\pi^2}\zeta(3)$$

Examples

(1) We have by (4.38) and the preceding theorem

$$\sum_{n\geq 1}^{\mathcal{R}} Log(2n-1) = \frac{1}{2}Log(2) - \frac{1}{2}$$

thus

$$\sum_{n\geq 1}^{\mathcal{R}} \left(Log(2n-1) - Log(2n) - \frac{1}{2n}\right) = Log(\sqrt{2\pi}) - \frac{1}{2}\gamma$$

and, since we are in a case of convergence, we get

$$\sum_{n=1}^{+\infty} \left(Log(2n-1) - Log(2n) - \frac{1}{2n}\right) = Log(\sqrt{\pi}) - \frac{1}{2}\gamma + \frac{1}{2}$$

Note that for $f \in \mathcal{O}^{\pi/2}$, by the shift property applied to $x \mapsto f(2x-1)$, we have

$$\sum_{n\geq 1}^{\mathcal{R}} f(2n+1) = \frac{1}{2}\sum_{n\geq 1}^{\mathcal{R}} f(n) + \frac{1}{2}\sum_{n\geq 1}^{\mathcal{E}} (-1)^{n-1}f(n) - f(1) + \int_1^2 f(2t-1)dt$$

In the special case $f = Log$ this gives

$$\sum_{n\geq 1}^{\mathcal{R}} Log(2n+1) = \frac{1}{2}Log(2) + \frac{3}{2}Log(3) - \frac{3}{2}$$

Thus, if we consider the sum $\sum_{n\geq 1}^{\mathcal{R}}(Log(2n-1) + Log(2n+1))$, we get

$$\sum_{n\geq 1}^{\mathcal{R}} Log(4n^2-1) = \sum_{n\geq 1}^{\mathcal{R}} Log(2n-1) + \sum_{n\geq 1}^{\mathcal{R}} Log(2n+1) = Log(2) + \frac{3}{2}Log(3) - 2$$

which gives

$$\sum_{n\geq 1}^{\mathcal{R}} Log(1 - \frac{1}{4n^2}) = Log(2) + \frac{3}{2}Log(3) - 2 - \sum_{n\geq 1}^{\mathcal{R}} Log(4n^2) = \frac{3}{2}Log(3) - Log(2\pi)$$

Since we are in a case of convergence, we get

$$\sum_{n=1}^{+\infty} Log(1 - \frac{1}{4n^2}) = Log(\frac{2}{\pi})$$

and by expanding the logarithm this gives

$$\sum_{k=1}^{+\infty} \frac{\zeta(2k)}{4^k k} = Log(\frac{\pi}{2})$$

If we consider the difference $\sum_{n\geq 1}^{\mathcal{R}} Log(2n+1) - \sum_{n\geq 1}^{\mathcal{R}} Log(2n-1)$ we find

$$\sum_{n=1}^{+\infty} Log(\frac{2n+1}{2n-1}) - \frac{1}{n} = Log(2) - \gamma$$

and by the use of

$$Log(\frac{1 + \frac{1}{2x}}{1 - \frac{1}{2x}}) = \sum_{k=0}^{+\infty} \frac{1}{2^{2k}(2k+1)} \frac{1}{x^{2k+1}}$$

this gives

$$\sum_{k=1}^{+\infty} \frac{\zeta(2k+1)}{2^{2k}(2k+1)} = Log(2) - \gamma$$

(2) Since for $s \neq 1$ we have

$$\sum_{n\geq 1}^{\mathcal{E}} \frac{(-1)^{n-1}Log(n)}{n^s} = \zeta'(s)(1 - 2^{1-s}) - \zeta(s)2^{1-s}Log(2)$$

we get by the preceding theorem

$$\sum_{n\geq 1}^{\mathcal{R}}(2n-1)Log(2n-1) = \zeta'(-1) + \frac{1}{6}Log(2) - \frac{1}{8}$$

$$\sum_{n\geq 1}^{\mathcal{R}}(2n+1)Log(2n+1) = \zeta'(-1) + \frac{1}{6}Log(2) + \frac{9}{4}Log(3) - \frac{9}{8}$$

We deduce that

$$\sum_{n\geq 1}^{\mathcal{R}} nLog(4n^2 - 1) = \zeta'(-1) + \frac{1}{6}Log(2) + \frac{3}{8}Log(3) - \frac{1}{8}$$

This is related to the series

$$\sum_{k=1}^{+\infty} \frac{\zeta(2k+1)}{4^k(k+1)} = -\sum_{k=1}^{+\infty}(4nLog(4n^2 - 1) - 4nLog(4n^2) + \frac{1}{n})$$

and we obtain

$$\sum_{k=1}^{+\infty} \frac{\zeta(2k+1)}{4^k(k+1)} = -12\,\zeta'(-1) - \gamma - \frac{1}{3}Log(2) - 1$$

By the same type of calculations we have

$$\sum_{n\geq 1}^{\mathcal{R}}(2n - 1)^2 Log(2n - 1) = 3\zeta'(-2) - \frac{1}{18}$$

$$\sum_{n\geq 1}^{\mathcal{R}}(2n + 1)^2 Log(2n + 1) = 3\zeta'(-2) + \frac{9}{2}Log(3) - \frac{27}{18}$$

We deduce that

$$\sum_{n\geq 1}^{\mathcal{R}} 4n^2 Log(4n^2 - 1) = 6\zeta'(-2) + Log(2) + \frac{3}{2}Log(3) - \frac{14}{9}$$

This gives

$$\sum_{k=1}^{+\infty} \frac{\zeta(2k)}{4^k(k+1)} = -14\,\zeta'(-2) - Log(2) + \frac{1}{2}$$

(3) Consider an integer $k \geq 1$ and $f(x) = \frac{Log^k(x)}{x}$ then by the preceding theorem we easily get an expression of the Stieltjes constants

$$\gamma_k = \sum_{n\geq 1}^{\mathcal{R}} \frac{Log^k(n)}{n}$$

in terms of alternating series. We have

$$\sum_{n\geq 1}^{\mathcal{R}} \frac{Log^k(2n)}{2n} = \frac{1}{2}\sum_{n\geq 1}^{\mathcal{R}} \frac{Log^k(n)}{n} - \frac{1}{2}\sum_{n\geq 1}^{\mathcal{E}} \frac{(-1)^{n-1}Log^k(n)}{n} + \frac{1}{2}\int_1^2 \frac{Log^k(x)}{x}dx$$

$$= \frac{1}{2}\gamma_k - \frac{1}{2}\sum_{n\geq 1}^{\mathcal{E}} \frac{(-1)^{n-1}Log^k(n)}{n} + \frac{1}{2}\frac{Log^{k+1}(2)}{k+1}$$

but the binomial expansion gives

$$\sum_{n\geq 1}^{\mathcal{R}} \frac{Log^k(2n)}{2n} = \frac{1}{2}\sum_{n\geq 1}^{\mathcal{R}} \frac{(Log(2) + Log(n))^k}{n} = \frac{1}{2}\sum_{j=0}^{k} C_k^j Log^{k-j}(2)\sum_{n\geq 1}^{\mathcal{R}} \frac{Log^j(n)}{n}$$

thus (with $\gamma_0 = \gamma$) we get

$$\sum_{n\geq 1}^{\mathcal{R}} \frac{Log^k(2n)}{2n} = \frac{1}{2}\sum_{j=0}^{k} C_k^j \gamma_j Log^{k-j}(2)$$

Since we are in a case of convergence for the series $\sum_{n\geq 1} \frac{(-1)^{n-1}Log^k(n)}{n}$ we obtain

$$\sum_{j=0}^{k-1} C_k^j \gamma_j Log^{k-j}(2) = \tilde{\gamma}_k \qquad (4.48)$$

where

$$\tilde{\gamma}_k = \frac{Log^{k+1}(2)}{k+1} - \sum_{n\geq 1}^{+\infty} \frac{(-1)^{n-1}Log^k(n)}{n}$$

We have for example

$$\gamma Log(2) = \frac{Log^2(2)}{2} - \sum_{n\geq 1}^{+\infty} \frac{(-1)^{n-1}Log(n)}{n}$$

$$\gamma Log^2(2) + 2\gamma_1 Log(2) = \frac{Log^3(2)}{3} - \sum_{n\geq 1}^{+\infty} \frac{(-1)^{n-1}Log^2(n)}{n}$$

$$\gamma Log^3(2) + 3\gamma_1 Log^2(2) + 3\gamma_2 Log(2) = \frac{Log^4(2)}{4} - \sum_{n\geq 1}^{+\infty} \frac{(-1)^{n-1}Log^3(n)}{n}$$

We can easily invert these relations by using the exponential generating functions

$$G(z) = \sum_{k \geq 0} \gamma_k \frac{z^k}{k!} \text{ and } \tilde{G}(z) = \sum_{k \geq 1} \tilde{\gamma}_k \frac{z^k}{k!}$$

The relation (4.48) is simply

$$(e^{z Log(2)} - 1)G(z) = \tilde{G}(z)$$

thus we have

$$G(z) = \frac{1}{e^{z Log(2)} - 1} \tilde{G}(z)$$

This gives Stieltjes constants γ_k in terms of linear combinations of the constants $\tilde{\gamma}_k$ involving powers of $Log(2)$ and Bernoulli numbers (Zhang and Williams 1994).

Remark Let $g \in \mathcal{O}^{\pi/4}$, then by applying the formulas of the preceding theorem to the function $x \mapsto g(2x + 1)$, we also get

$$\sum_{n \geq 1}^{\mathcal{R}} g(4n - 1) = \frac{1}{2} \sum_{n \geq 1}^{\mathcal{R}} g(2n + 1) + \frac{1}{2} \sum_{n \geq 1}^{\mathcal{E}} (-1)^{n-1} g(2n + 1)$$

$$\sum_{n \geq 1}^{\mathcal{R}} g(4n + 1) = \frac{1}{2} \sum_{n \geq 1}^{\mathcal{R}} g(2n + 1) - \frac{1}{2} \sum_{n \geq 1}^{\mathcal{E}} (-1)^{n-1} g(2n + 1) + \frac{1}{2} \int_1^2 g(2t + 1) dt$$

4.3.4 Generalization

Let N be an integer > 1 and

$$\Omega_N = \{e^{2i\pi m/N}, m = 1, \ldots, N - 1\}$$

For a root of unity $\omega \in \Omega_N$ and a function f in $\mathcal{O}^{2\pi/N}$ we can define a sort of Euler summation of the series $\sum_{n \geq 1} \omega^{n-1} f(n)$ by

$$\sum_{n \geq 1}^{\mathcal{E}_\omega} \omega^{n-1} f(n) = A_f^\omega(1)$$

where A_f^ω is the solution in $\mathcal{O}^{2\pi/N}$ of the equation

$$A(x) - \omega A(x + 1) = f(x)$$

Like in Lemma 19 this unique solution $A_f^\omega \in \mathcal{O}^{2\pi/N}$ is given by

$$A_f^\omega(x) = \sum_{k=0}^{N-1} \omega^k R_{f(Nx)}\left(\frac{x+k}{N}\right) \tag{4.49}$$

(where we use the notation $f(Nx)$ for the function $: x \mapsto f(Nx)$).

Note that if the series $\sum_{n\geq 1} \omega^{n-1} f(n+x-1)$ is convergent for $Re(x) > 0$ then we immediately have

$$A_f^\omega(x) = \sum_{n\geq 1}^{+\infty} \omega^{n-1} f(n+x-1)$$

thus in this case

$$\sum_{n\geq 1}^{\mathcal{E}_\omega} \omega^{n-1} f(n) = \sum_{n\geq 1}^{+\infty} \omega^{n-1} f(n)$$

The relation with the Ramanujan summation is easily obtained by (4.49) since, for $k = 0, \ldots, N-1$, we have by (2.2) (with $x + 1 = \frac{k+1}{N}$)

$$R_{f(Nx)}\left(\frac{k+1}{N}\right) = \sum_{n\geq 1}^{\mathcal{R}} f(Nn - N + 1 + k) - \int_1^{\frac{1+k}{N}} f(Nx)dx$$

thus we deduce from (4.49) that, for $\omega \in \Omega_N$, we have

$$\sum_{n\geq 1}^{\mathcal{E}_\omega} \omega^{n-1} f(n) = \sum_{k=0}^{N-1} \omega^k \sum_{n\geq 1}^{\mathcal{R}} f(Nn + k + 1 - N) + \frac{1}{N} \sum_{k=0}^{N-1} \omega^k \int_{k+1}^N f(x)dx$$

Since $\omega \in \Omega_N = \{e^{2i\pi m/N}, m = 1, \ldots, N-1\}$, this gives $N-1$ equations. There is a supplementary equation (using (2.16) with f replaced by $f(Nx)$) that is

$$\sum_{n\geq 1}^{\mathcal{R}} f(n) + \int_1^N f(x)dx = \sum_{k=0}^{N-1} \sum_{n\geq 1}^{\mathcal{R}} f(Nn + k + 1 - N) + \frac{1}{N} \sum_{k=0}^{N-1} \int_{k+1}^N f(x)dx$$

Thus we have a system of N equations of type

$$a_m = \sum_{k=0}^{N-1} b_k e^{\frac{2i\pi m}{N} k} \quad \text{with } m = 0, \ldots, N-1$$

where

$$b_k = \sum_{n\geq 1}^{\mathcal{R}} f(Nn + k + 1 - N) + \frac{1}{N}\int_{k+1}^{N} f(x)dx$$

Since such a system can be solved by

$$b_k = \frac{1}{N}\sum_{m=0}^{N-1} e^{-\frac{2i\pi m}{N}k} a_m$$

we deduce that, for $k = 0, \ldots, N-1$, we have

$$\sum_{n\geq 1}^{\mathcal{R}} f(Nn + k + 1 - N) = \frac{1}{N}\sum_{n\geq 1}^{\mathcal{R}} f(n) + \frac{1}{N}\sum_{\omega\in\Omega_N} \omega^{-k}\sum_{n\geq 1}^{\mathcal{E}_\omega} \omega^{n-1}f(n)$$

$$+ \frac{1}{N}\int_1^{k+1} f(x)dx \ .$$

Conclusion

As in Theorem 20 we have, for $k = 1, \ldots, N$, the following relation

$$\sum_{n\geq 1}^{\mathcal{R}} f(Nn + k - N) = \frac{1}{N}\sum_{n\geq 1}^{\mathcal{R}} f(n) + \frac{1}{N}\sum_{\omega\in\Omega_N} \omega^{-k+1}S_f(\omega) + \frac{1}{N}\int_1^k f(x)dx \qquad (4.50)$$

where

$$S_f(\omega) = \sum_{n\geq 1}^{\mathcal{E}_\omega} \omega^{n-1}f(n)$$

For example with $N = 3$ we have

$$\sum_{n\geq 1}^{\mathcal{R}} f(3n - 2) = \frac{1}{3}\sum_{n\geq 1}^{\mathcal{E}} f(n) + \frac{1}{3}S_f(e^{\frac{2i\pi}{3}}) + \frac{1}{3}S_f(e^{\frac{4i\pi}{3}})$$

$$\sum_{n\geq 1}^{\mathcal{R}} f(3n - 1) = \frac{1}{3}\sum_{n\geq 1}^{\mathcal{E}} f(n) + \frac{1}{3}e^{-\frac{2i\pi}{3}}S_f(e^{\frac{2i\pi}{3}}) + \frac{1}{3}e^{-\frac{4i\pi}{3}}S_f(e^{\frac{4i\pi}{3}})$$

$$+ \frac{1}{3}\int_1^2 f(x)dx$$

$$\overset{\mathcal{R}}{\underset{n\geq 1}{\sum}} f(3n) = \frac{1}{3} \overset{\mathcal{E}}{\underset{n\geq 1}{\sum}} f(n) + \frac{1}{3} e^{-\frac{4i\pi}{3}} S_f(e^{\frac{2i\pi}{3}}) + \frac{1}{3} e^{-\frac{2i\pi}{3}} S_f(e^{\frac{4i\pi}{3}})$$

$$+ \frac{1}{3} \int_1^3 f(x)dx$$

Remark For $k = N$ we have by (4.50)

$$N \overset{\mathcal{R}}{\underset{n\geq 1}{\sum}} f(Nn) = \overset{\mathcal{R}}{\underset{n\geq 1}{\sum}} f(n) + \underset{\omega \in \Omega_N}{\sum} \omega S_f(\omega) + \int_1^N f(x)dx \qquad (4.51)$$

If we are in a case of convergence for the series $\sum_{n\geq 1} \omega^{n-1} f(n)$ then

$$\omega S_f(\omega) = \overset{+\infty}{\underset{n=1}{\sum}} \omega^n f(n)$$

Therefore (4.51) is

$$N \overset{\mathcal{R}}{\underset{n\geq 1}{\sum}} f(Nn) = \overset{\mathcal{R}}{\underset{n\geq 1}{\sum}} f(n) + \overset{+\infty}{\underset{n=1}{\sum}} \underset{\omega \in \Omega_N}{\sum} \omega^n f(n) + \int_1^N f(x)dx$$

Thus for every integer $N > 1$ we have

$$N \overset{\mathcal{R}}{\underset{n\geq 1}{\sum}} f(Nn) = \overset{\mathcal{R}}{\underset{n\geq 1}{\sum}} f(n) + \overset{+\infty}{\underset{n=1}{\sum}} \varepsilon_N(n) f(n) + \int_1^N f(x)dx \qquad (4.52)$$

where

$$\varepsilon_N(n) = \underset{\omega \in \Omega_N}{\sum} \omega^n = N - 1 \text{ if } N|n$$

$$= -1 \quad \text{if } N \nmid n$$

Applying this formula for $f(x) = \frac{Log(x)}{x}$, we obtain for every integer $N > 1$

$$\gamma = \frac{1}{Log(N)} \overset{+\infty}{\underset{n=1}{\sum}} \varepsilon_N(n) \frac{Log(n)}{n} + \frac{1}{2} Log(N)$$

and for $f(x) = \frac{Log^2(x)}{x}$, we have

$$\gamma_1 = \frac{1}{2Log(N)} \sum_{n=1}^{+\infty} \varepsilon_N(n) \frac{Log^2(n)}{n} + \frac{1}{6} Log^2(N) - \frac{1}{2} \gamma Log(N)$$

For example with $N = 3$, we have $\sum_{\omega \in \Omega_3} \omega^n = 2 \cos(\frac{2n\pi}{3})$, and we get

$$\gamma = \frac{2}{Log(3)} \sum_{n=1}^{+\infty} \cos(\frac{2n\pi}{3}) \frac{Log(n)}{n} + \frac{1}{2} Log(3)$$

$$\gamma_1 = \frac{1}{Log(3)} \sum_{n=1}^{+\infty} \cos(\frac{2n\pi}{3}) \frac{Log^2(n)}{n} + \frac{1}{6} Log^2(3) - \frac{1}{2} \gamma Log(3)$$

Chapter 5
An Algebraic View on the Summation of Series

The Ramanujan summation differs from the classical summation methods by the fact that for convergent series it does not give the usual sum. Also there is the shift property which seems very strange for a summation procedure. Thus it is necessary to define a general algebraic formalism to unify the Ramanujan summation and the classical methods of summation of series.

5.1 Introduction

To introduce this formalism we begin with the analysis of the Borel summation. We have seen in (4.1.2) that the Borel summation is formally given by the formula

$$\sum_{n\geq 0}^{\mathcal{B}} a_n = \int_0^{+\infty} e^{-t} \left(\sum_{n=0}^{+\infty} a_n \frac{t^n}{n!} \right) dt$$

We now show that this formula is simply related to the resolution of a differential equation. More precisely let us consider a complex sequence (a_n), such that the series

$$f(x) = \sum_{n=0}^{+\infty} a_n \frac{x^n}{n!}$$

is convergent for x near 0, then the function f is analytic near 0 and such that for all $n \geq 0$ we have

$$a_n = \partial^n f(0)$$

© Springer International Publishing AG 2017
B. Candelpergher, *Ramanujan Summation of Divergent Series*,
Lecture Notes in Mathematics 2185, DOI 10.1007/978-3-319-63630-6_5

And formally we get

$$\sum_{n\geq 0} a_n = \sum_{n\geq 0} \partial^n f(0) = (\sum_{n\geq 0} \partial^n f)(0) = ((I - \partial)^{-1} f)(0)$$

Thus $\sum_{n\geq 0} a_n = R(0)$, where the function R is a solution of the differential equation

$$(I - \partial)R = f$$

Assume that f has an analytic continuation near $[0, +\infty[$, then this equation has the general solution

$$R(x) = -e^x \int_0^x e^{-t} f(t)dt + Ke^x \text{ with } K \in \mathbb{C}$$

To select one solution of this equation we must set a condition on R. We set the condition

$$\lim_{x\to +\infty} e^{-x} R(x) = 0$$

This condition is equivalent to the convergence of the integral $\int_0^{+\infty} e^{-t} f(t)dt$ with

$$K = \int_0^{+\infty} e^{-t} f(t)dt$$

and gives the unique solution

$$R(x) = e^x \int_x^{+\infty} e^{-t} f(t)dt$$

and finally we get formally

$$\sum_{n\geq 0} a_n = R(0) = \int_0^{+\infty} e^{-t} f(t)dt$$

This suggests the following presentation of the Borel summation:

Let E be the space of complex analytic functions f on $[0, +\infty[$ such that for all $n \geq 0$ the function $x \mapsto e^{-x} \partial^n f(x)$ has a finite limit when $x \to +\infty$.

And let's define the operators

$$D(f)(x) = \partial f(x)$$
$$v_0(f) = f(0)$$
$$v_\infty(f) = \lim_{x\to +\infty} e^{-x} f(x)$$

We say that a sequence (a_n) *has the generating function* $f \in E$ if

$$a_n = \partial^n f(0)$$

Since f is analytic near 0, then in a small disk $D(0, \rho)$ we have

$$f(x) = \sum_{n=0}^{+\infty} \partial^n f(0) \frac{x^n}{n!}$$

Then, by the preceding calculations, the differential equation

$$R - \partial R = f$$
$$v_\infty(f) = 0$$

gives a unique solution R_f given by

$$R(x) = e^x \int_x^{+\infty} e^{-t} f(t) dt$$

And we say that the series $\sum_{n \geq 0} a_n$ is Borel summable if the series

$$\sum_{n \geq 0} a_n \frac{x^n}{n!}$$

is convergent for x near 0 and defines by analytic continuation a function $f \in E$ and we have

$$\sum_{n \geq 0}^{\mathcal{B}} a_n = v_0(R_f)$$

5.2   An Algebraic Formalism

Definition 7 A summation space $\mathcal{T} = (E, D, v_0, v_\infty)$ is given by a $\mathbb{C}$-vector space E with a linear operator $D : E \to E$ and two linear "evaluation operators" $v_0 : E \to \mathbb{C}$ and $v_\infty : E \to \mathbb{C}$ such that:

(*) The solutions of the equation

$$Dg = g$$

form a one dimensional subspace of E generated by an element $\alpha \in E$ with

$$v_0(\alpha) = v_\infty(\alpha) = 1$$

(**) If $g \in E$ is such that $v_0(D^n g) = 0$ for all $n \geq 0$ then $g = 0$

Remark By the property (*) we deduce that if $Dg = g$ and $v_\infty(g) = 0$ then $g = 0$.

Definition 8 Let (a_n) be a complex sequence, we say that this sequence *is generated by* $f \in E$ if

$$a_n = v_0(D^n f) \text{ for all } n \geq 0$$

Then by (**) this element $f \in E$ is unique and is *the generator* of the sequence (a_n).

The constant sequence defined by $a_n = 1$ for every $n \geq 1$ is generated by the element α since $D^n \alpha = \alpha$ and $v_0(\alpha) = 1$.

Note that if (a_n) is generated by f and (b_n) is generated by g, then for any complex numbers C, D the sequence $(Ca_n + Db_n)$ is generated by $Cf + Dg$.

Summation of a Series

To define the sum $\sum_{n \geq 0} a_n$ we can write formally

$$\sum_{n \geq 0} a_n = \sum_{n \geq 0} v_0(D^n f) = v_0\left(\sum_{n \geq 0} D^n f\right) = v_0((I - D)^{-1} f)$$

thus we get

$$\sum_{n \geq 0} a_n = v_0(R)$$

where R is a solution of the equation

$$(I - D)R = f$$

To have the uniqueness of the sum it suffices to have uniqueness of the solution R of this equation. By linearity of D this is equivalent to say that:

$$\text{if } DR = R \text{ then } R = 0$$

by the preceding remark this is the case if we add the condition

$$v_\infty(R) = 0$$

Definition 9 Consider a summation space $\mathcal{T} = (E, D, v_0, v_\infty)$. Let (a_n) be a complex sequence generated by an element $f \in E$ and let's assume that there exists $R_f \in E$ which satisfies

$$R_f - DR_f = f \text{ with } v_\infty(R_f) = 0 \tag{5.1}$$

Then we say that $\sum_{n\geq 0} a_n$ is $\mathcal{T}$-summable and we define the sum $\sum_{n\geq 0}^{\mathcal{T}} a_n$ by

$$\sum_{n\geq 0}^{\mathcal{T}} a_n = v_0(R_f)$$

Examples

(1) The usual Cauchy summation

Let E be the vector space of convergent complex sequences $u = (u_n)_{n\geq 0}$. Let's define the operators

$$D : (u_0, u_1, u_2, \ldots) \mapsto (u_1, u_2, u_3, \ldots)$$

$$v_0 : (u_0, u_1, u_2, \ldots) \mapsto u_0$$

$$v_\infty : (u_0, u_1, u_2, \ldots) \mapsto \lim_{n\to+\infty} u_n$$

We have the two properties:

(*) The solutions of the equation $Dg = g$ form the one-dimensional subspace of E generated by the element

$$\alpha = (1, 1, 1, \ldots)$$

(**) If $v_0(D^n g) = 0$ for all $n \geq 0$, then $g = 0$ since $v_0(D^n g) = g_n$ for every $g \in E$.

Let $\mathcal{C}$ be this summation space. A complex sequence (a_n) is generated by $f \in E$ if $v_0(D^n f) = a_n$ and since $v_0(D^n f) = f_n$ we see that every complex sequence (a_n) has the generating element

$$f = (a_n)$$

To define $\sum_{n\geq 0}^{\mathcal{C}} a_n$ we must solve the equation

$$R - DR = f \qquad (5.2)$$

that is

$$(R_0, R_1, R_2, \ldots) - (R_1, R_2, R_3, \ldots) = (a_0, a_1, a_2, \ldots)$$

this gives

$$R_n - R_{n+1} = a_n$$

thus we have

$$R_{n+1} = R_0 - \sum_{k=0}^{n} a_k$$

This gives an infinity of solutions of (5.2) since R_0 is undetermined, thus we add to (5.2) the condition $v_\infty(R) = 0$, that is

$$\lim_{n \to +\infty} R_{n+1} = 0$$

which is equivalent to say that $\sum_{k=0}^{n} a_k$ has a finite limit when $n \to +\infty$ and

$$\lim_{n \to +\infty} \sum_{k=0}^{n} a_k = R_0 = v_0(R)$$

Finally we see that the series $\sum_{n \geq 0} a_n$ is Cauchy-summable if $\sum_{k=0}^{n} a_k$ has a finite limit when $n \to +\infty$, in this case we say that the series $\sum_{n \geq 0} a_n$ *is convergent* and we write simply

$$\sum_{n \geq 0}^{\mathcal{C}} a_n = \sum_{n=0}^{+\infty} a_n = \lim_{n \to +\infty} \sum_{k=0}^{n} a_k$$

that is the usual sum of a convergent series.

(2) The Ramanujan summation

First note that in the preceding chapters we have defined the Ramanujan summation for series $\sum_{n \geq 1} f(n)$ indexed by $n \geq 1$, these can be seen as series indexed by $n \geq 0$ if we set

$$\sum_{n \geq 1} f(n) = \sum_{n \geq 0} a_n \text{ where } a_n = f(n+1)$$

Consider the space $E = \mathcal{O}^\pi$ and the operators

$$Df(x) = f(x+1)$$
$$v_0(f) = f(1)$$
$$v_\infty(f) = \int_1^2 f(t) dt$$

We have the two properties:

(*) The solutions of the equation $Dg = g$ are the functions $g \in \mathcal{O}^\pi$ such that

$$g(x+1) = g(x)$$

This gives $g(n) = g(1)$ for every integer $n \geq 1$ and by Carlson's theorem this implies that g is a constant function, thus g is in the one-dimensional subspace of E generated by the constant function $\alpha = 1$.

(**) If $v_0(D^n g) = 0$ for all $n \geq 0$, then by Carlson's theorem $g = 0$ since $v_0(D^n g) = g(n)$.

Consider $\mathcal{R}$ this summation space. A complex sequence $(a_n)_{n \geq 0}$ is generated by an element $f \in \mathcal{O}^\pi$ if for all integer $n \geq 0$ we have

$$a_n = v_0(D^n f) = f(n+1)$$

The equation

$$R_f - DR_f = f$$

is in this case our usual difference equation

$$R_f(x) - R_f(x+1) = f(x)$$

and the condition

$$v_\infty(R_f) = 0$$

is simply the condition

$$\int_1^2 R_f(t)dt = 0$$

that we have used in the Ramanujan summation. And we have

$$\sum_{n \geq 0}^{\mathcal{R}} a_n = v_0(R_f) = R_f(1)$$

which is the definition used in the preceding chapters.

5.3   Properties of the General Summation

5.3.1   The Linearity Property

For any complex numbers C, D we have

$$\sum_{n \geq 0}^{\mathcal{T}} C a_n + D b_n = C \sum_{n \geq 0}^{\mathcal{T}} a_n + D \sum_{n \geq 0}^{\mathcal{T}} b_n \tag{5.3}$$

which is an immediate consequence of linearity of D, v_0, v_∞.

5.3.2   The Shift Property

Theorem 21 *If the sequence (a_n) is generated by $f \in E$, then for any integer $N \geq 1$ we have the shift property*

$$\sum_{n\geq 0}^{\mathcal{T}} a_{n+N} = \sum_{n\geq 0}^{\mathcal{T}} a_n - \sum_{n=0}^{N-1} a_n + \sum_{k=0}^{N-1} v_\infty(D^k f) \qquad (5.4)$$

In the special case $N = 1$ we get

$$\sum_{n\geq 0}^{\mathcal{T}} a_{n+1} = \sum_{n\geq 0}^{\mathcal{T}} a_n - a_0 + v_\infty(f) \qquad (5.5)$$

Proof First of all we must prove that the series $\sum_{n\geq 0} a_{n+N}$ is $\mathcal{T}$-summable.
If (a_n) is generated by $f \in E$ then for any integer $N \geq 1$ we have

$$a_{n+N} = v_0(D^{n+N} f) = v_0(D^n(D^N f))$$

thus the sequence (a_{n+N}) is generated by $D^N f \in E$. Then the equation $R_f - DR_f = f$ gives

$$D^N R_f - D(D^N R_f) = D^N f$$

but generally we don't have $v_\infty(D^N R_f) = 0$. Thus we consider

$$T_N = D^N R_f - v_\infty(D^N R_f)\alpha$$

then we have immediately

$$T_N - DT_N = D^N f$$
$$v_\infty(T_N) = 0$$

Thus by Definition 9 we have the $\mathcal{T}$-summability of $\sum_{n\geq 0} a_{n+N}$ and

$$\sum_{n\geq 0}^{\mathcal{T}} a_{n+N} = v_0(T_N) = v_0(D^N(R_f)) - v_\infty(D^N R_f) \qquad (5.6)$$

Let us now evaluate the expression $\sum_{n\geq 0}^{\mathcal{T}} a_n - \sum_{n=0}^{N-1} a_n$ in the right side of Eq. (5.4).
By summation for k from 0 to $N - 1$ ($N \geq 1$) of the relations

$$D^k R_f - D^{k+1} R_f = D^k f$$

we get

$$R_f - D^N R_f = \sum_{k=0}^{N-1} D^k f$$

Since $v_0(D^k f) = a_k$, this gives

$$\sum_{n\geq 0}^{\mathcal{T}} a_n - \sum_{k=0}^{N-1} a_k = v_0(D^N R_f) \tag{5.7}$$

And by (5.6) we obtain

$$\sum_{n\geq 0}^{\mathcal{T}} a_{n+N} = \sum_{n\geq 0}^{\mathcal{T}} a_n - \sum_{n=0}^{N-1} a_n - v_\infty(D^N R_f) \tag{5.8}$$

To get (5.4) it suffices to note that

$$v_\infty(D^N R_f) = v_\infty(R_f - \sum_{k=0}^{N-1} D^k f) = -\sum_{k=0}^{N-1} v_\infty(D^k f)$$

□

Remark Note that if in Definition 7 we set the additional property:

(***) If $v_\infty(g) = 0$ then $v_\infty(Dg) = 0$

then $v_\infty(R_f) = 0$ gives $v_\infty(D^N R_f) = 0$ for all positive integer N, thus (5.8) gives the usual property

$$\sum_{n\geq 0}^{\mathcal{T}} a_{n+N} = \sum_{n\geq 0}^{\mathcal{T}} a_n - \sum_{n=0}^{N-1} a_n$$

This property is verified for most summations but not for the Ramanujan summation.

5.3.3 The Associated Algebraic Limit

To define an algebraic notion of limit in a summation space $\mathcal{T}$ we start with the following simple remark concerning the case of the usual Cauchy summation: consider a complex sequence $(a_n)_{n\geq 0}$, then for every integer $N \geq 0$ we have

$$\sum_{n\geq 0}^{N} (a_n - a_{n+1}) = a_0 - a_{N+1}$$

thus the series $\sum_{n\geq 0}^{N}(a_n - a_{n+1})$ is convergent if and only if $\lim_{n\to+\infty} a_n$ exists and we have

$$\lim_{n\to+\infty} a_n = a_0 - \sum_{n\geq 0}^{+\infty}(a_n - a_{n+1})$$

Similarly if the series $\sum_{n\geq 0} a_n$ is $\mathcal{T}$-summable it is natural to *define the generalized limit of the sequence* (a_n) by

$$\lim_{\mathcal{T}} a_n = a_0 - \sum_{n\geq 0}^{\mathcal{T}}(a_n - a_{n+1})$$

Then by the shift property (5.5) we see that the $\mathcal{T}$-limit of a sequence (a_n) with generator $f \in E$ is simply

$$\lim_{\mathcal{T}} a_n = v_\infty(f)$$

Now it remains to prove that like the Cauchy-summation the $\mathcal{T}$-summation is related to the $\mathcal{T}$-limit of the sequence of *partial sums* defined by

$$s_0(a) = 0 \text{ and } s_n(a) = \sum_{k=0}^{n-1} a_k = 0 \text{ for } n \geq 1 \tag{5.9}$$

If the series $\sum_{n\geq 0} a_n$ is $\mathcal{T}$-summable we define the sequence $(r_n(a))$ *of remainders of this series* by

$$r_0(a) = \sum_{n\geq 0}^{\mathcal{T}} a_n \text{ and } r_n(a) = r_0(a) - \sum_{k=0}^{n-1} a_k \text{ for } n \geq 1 \tag{5.10}$$

Since we have $r_0(a) = v_0(R_f)$ and by (5.7) for any integer $n \geq 1$

$$r_n(a) = v_0(D^n R_f) \tag{5.11}$$

therefore we see that the sequence $(r_n(a))$ is generated by R_f. Thus we have

$$\lim_{\mathcal{T}} r_n(a) = v_\infty(R_f) = 0$$

Since $s_n(a) = r_0(a) - r_n(a)$, then the sequence $(s_n(a))$ is generated by the element

$$r_0(a)\alpha - R_f \in E$$

and we have

$$\lim_{\mathcal{T}}(\sum_{k=0}^{n-1} a_k) = v_\infty(r_0(a)\alpha - R_f) = r_0(a) = \sum_{n\geq 0}^{\mathcal{T}} a_n$$

Remark In the case of the Ramanujan summation we have seen that a sequence $(a_n)_{n\geq 0}$ is generated by $f \in \mathcal{O}^\pi$ if $a_n = f(n+1)$ for every integer $n \geq 0$. Thus

$$s_n(a) = \sum_{k=0}^{n-1} a_k = \sum_{k=0}^{n-1} f(k+1) = \sum_{k=1}^{n} f(k) = \varphi_f(n)$$

Since we have

$$\varphi_f(n) = \varphi_f(n+1) - f(n+1)$$

we see that the generating function of the sequence $(s_n(a))$ is $\varphi_f - f$.

Thus in the general case it is natural to define φ_f by

$$\varphi_f = r_0(a)\alpha - R_f + f$$

it is the generating element of the sequence $(s_{n+1}(a) = \sum_{k=0}^{n} a_k)$. Note that

$$\lim_{\mathcal{T}}(\sum_{k=0}^{n} a_k) = v_\infty(\varphi_f) = \sum_{n\geq 0}^{\mathcal{T}} a_n + v_\infty(f)$$

5.3.4 Sum of Products

We suppose now that for certain elements f, g in the $\mathcal{T}$-summation space E we can define a product

$$(f, g) \mapsto f.g \in E$$

which has the usual properties of associativity, commutativity and distributivity. And we suppose that for the evaluation operator v_0 we have

$$v_0(f.g) = v_0(f)v_0(g)$$

Suppose that for f, g in E we have

$$D(f.g) = Df.Dg \tag{5.12}$$

by induction we get for $n \geq 1$

$$D^n(f.g) = (D^n f).(D^n g)$$

If (a_n) is generated by f and (b_n) is generated by g then the sequence $(a_n b_n)$ is generated by $f.g$, since

$$a_n b_n = v_0(D^n f) v_0(D^n g) = v_0((D^n f).(D^n g)) = v_0(D^n(f.g))$$

To get a formula for $\sum_{n \geq 0}^{\mathcal{T}} a_n b_n$ we observe that

$$R_f.R_g - D(R_f.R_g) = R_f.R_g - D(R_f).D(R_g)$$
$$= (R_f - D(R_f)).R_g + D(R_f).(R_g - D(R_g))$$
$$= f.R_g + g.R_f - f.g$$

thus if $R_f.R_g, f.R_g$, and $g.R_f$ are in E we get

$$R_{f.R_g + g.R_f - f.g} = R_f.R_g - v_\infty(R_f.R_g) \tag{5.13}$$

Since R_f is the generator of the sequence of remainders $(r_n(a))$ and R_g is the generator of $(r_n(b))$ then by (5.10) we get

$$\sum_{n \geq 0}^{\mathcal{T}} a_n b_n = \sum_{n \geq 0}^{\mathcal{T}} a_n r_n(b) + \sum_{n \geq 0}^{\mathcal{T}} b_n r_n(a) - \sum_{n \geq 0}^{\mathcal{T}} a_n \sum_{n \geq 0}^{\mathcal{T}} b_n + v_\infty(R_f.R_g)$$

This is the formula we encounter in the proof of Theorem 2 since in the case of the Ramanujan summation we have $R_f(n + 1) = r_n(a)$ if $f(n + 1) = a_n$. Note that the additional term $v_\infty(R_f.R_g)$ disappears in the case of the Cauchy summation, since in this case

$$v_\infty(R_f.R_g) = v_\infty(R_f) v_\infty(R_g) = 0$$

Remark More generally we can define a *convolution product of the sequences* (a_n) *and* (b_n) by

$$(a \star b)_n = v_0(D^n(f.g))$$

that is the sequence with $f.g$ as generator.

For example if we suppose that for f and g in E we have

$$D(f.g) = (Df).g + f.(Dg)$$

then by induction we get for $n \geq 1$

$$D^n(f.g) = \sum_{k=0}^{n} C_n^k (D^k f).(D^{n-k} g)$$

In this case we get

$$(a \star b)_n = \sum_{k=0}^{n} C_n^k a_k b_{n-k}$$

5.4 Examples

(1) The Cesaro summation

Let E be the vector space of complex sequences $u = (u_n)_{n \geq 0}$ such that

$$\lim_{n \to +\infty} \frac{u_0 + \ldots + u_{n-1}}{n} \text{ is finite}$$

And let's consider the operators

$$D : (u_n) \mapsto (u_1, u_2, u_3, \ldots)$$

$$v_0 : (u_n) \mapsto u_0$$

$$v_\infty : (u_n) \mapsto \lim_{n \to +\infty} \frac{u_0 + \ldots + u_{n-1}}{n}$$

A sequence $(a_n) \in E$ is generated by

$$f = (a_n)$$

The equation $R - DR = f$ is $R_k - R_{k+1} = a_k$, thus

$$R_0 - R_n = \sum_{k=0}^{n-1} a_k = s_n$$

By taking the sum of the equations

$$R_0 - R_1 = s_1$$
$$R_0 - R_2 = s_2$$
$$\ldots$$
$$R_0 - R_n = s_n$$

we get

$$R_0 - \frac{R_1 + \ldots + R_n}{n} = \frac{s_1 + \ldots + s_n}{n}$$

Thus $R = (R_n) \in E$ if and only if $\frac{s_1 + \ldots + s_n}{n}$ has a finite limit when $n \to +\infty$. The condition $v_\infty(R) = 0$ is equivalent to $\lim_{n \to +\infty} \frac{R_1 + \ldots + R_n}{n} = 0$, this gives

$$R_0 = \lim_{n \to +\infty} \frac{s_1 + \ldots + s_n}{n}$$

Thus the series $\sum_{n \geq 0} a_n$ is Cesaro-summable if the sequence $(\frac{s_1 + \ldots + s_n}{n})$ has a finite limit when $n \to +\infty$, and we write

$$\sum_{n=0}^{\mathcal{C}} a_n = v_0(R) = R_0 = \lim_{n \to +\infty} \frac{s_1 + \ldots + s_n}{n}$$

(2) The Euler summation

Let E be the vector space of complex sequences $u = (u_n)_{n \geq 0}$ such that

$$\lim_{n \to +\infty} \frac{u_n}{2^n} \text{ is finite}$$

And let's consider the operators

$$D : (u_n) \mapsto (u_{n+1} - u_n)$$

$$v_0 : (u_n) \mapsto u_0$$

$$v_\infty : (u_n) \mapsto \lim_{n \to +\infty} \frac{u_n}{2^n}$$

Let $f = (f_n)_{n \geq 0}$, we have for all $n \geq 0$

$$v_0(D^n f) = \sum_{k=0}^{n} C_n^k f_k (-1)^{n-k}$$

We deduce that a complex sequence (a_n) is generated by f if and only if

$$a_n = \sum_{k=0}^{n} C_n^k f_k (-1)^{n-k}$$

and by inversion of this relation we get

$$f_n = \sum_{k=0}^{n} a_k C_n^k$$

The equation $R - DR = f$ is

$$2R_k - R_{k+1} = f_k$$

this gives

$$R_0 = \frac{1}{2}R_1 + \frac{1}{2}f_0$$

$$\frac{1}{2}R_1 = \frac{1}{2^2}R_2 + \frac{1}{2^2}f_1$$

$$\frac{1}{2^2}R_2 = \frac{1}{2^3}R_3 + \frac{1}{2^3}f_2$$

$$\cdots$$

By taking the sum of these equations we get

$$R_0 = \frac{1}{2}f_0 + \frac{1}{2^2}f_1 + \ldots + \frac{1}{2^{n+1}}f_n - \frac{1}{2^n}R_n$$

We have $R \in E$ if and only if the sequence $(\frac{R_n}{2^n})$ has a finite limit, which is equivalent to say that

$$\sum_{n\geq 0} \frac{1}{2^{n+1}}f_n \text{ is convergent}$$

The condition

$$0 = v_\infty(R) = \lim_{n\to+\infty} \frac{1}{2^n}R_n$$

gives

$$R_0 = \lim_{n\to+\infty} \frac{1}{2}f_0 + \frac{1}{2^2}f_1 + \ldots + \frac{1}{2^{n+1}}f_n$$

In conclusion we see that the series $\sum_{n\geq 0} a_n$ is Euler summable if the series

$$\sum_{k\geq 0} \frac{1}{2^{n+1}} \left(\sum_{k=0}^{n} a_k C_n^k \right)$$

is convergent and in this case we have

$$\sum_{n=0}^{\varepsilon} a_n = \sum_{k=0}^{+\infty} \frac{1}{2^{n+1}} \left(\sum_{k=0}^{n} a_k C_n^k \right)$$

(3) **The Abel summation**

Let E be the vector space of analytic functions on $]-1,1[$ such that

$$\lim_{x \to 1} (1-x)f(x) \text{ is finite}$$

Let's consider the operators

$$Df(x) = \frac{f(x)-f(0)}{x} \text{ if } x \neq 0 \text{ and } Df(0) = f'(0)$$

$$v_0(f) = f(0)$$

$$v_\infty(f) = \lim_{x \to 1}(1-x)f(x)$$

Since $f \in E$ is analytic on $]-1,1[$ we can write

$$f(x) = \sum_{m=0}^{+\infty} \alpha_m x^m \text{ with } \alpha_m = \frac{\partial^m f(0)}{m!}$$

and we have

$$Df(x) = \sum_{m=0}^{+\infty} \alpha_{m+1} x^m$$

By induction we get for all $n \geq 0$

$$D^n f(x) = \sum_{m=0}^{+\infty} \alpha_{m+n} x^m$$

thus

$$v_0(D^n f) = \alpha_n = \frac{\partial^n f(0)}{n!}$$

Let's consider a complex sequence (a_n) and assume that the series

$$f(x) = \sum_{n=0}^{+\infty} a_n x^n$$

is convergent for $x \in]-1,1[$ and defines a function $f \in E$.

Then

$$a_n = \frac{\partial^n f(0)}{n!} = v_0(D^n f)$$

and the sequence (a_n) is generated by f.

The equation $R - DR = f$ is

$$R(x) - \frac{R(x) - R(0)}{x} = f(x) \text{ if } x \neq 0 \tag{5.14}$$

$$R(0) - R'(0) = f(0) \tag{5.15}$$

By (5.14) this gives

$$R(x) = \frac{1}{1-x}(R(0) - xf(x)) \text{ if } x \neq 0$$

Thus R is analytic on $]-1, 1[$ and (5.15) is automatically verified. We see that $R \in E$ if and only if $\lim_{x \to 1} f(x)$ is finite and the condition $v_\infty(R) = 0$ gives

$$R(0) = \lim_{x \to 1} f(x)$$

In conclusion we see that if the series $\sum_{n \geq 0} a_n x^n$ is convergent for all $x \in [-1, 1[$ and, if $\lim_{x \to 1} \sum_{n=0}^{+\infty} a_n x^n$ is finite, then $f \in E$ and $\sum_{n \geq 0} a_n$ is Abel-summable

$$\sum_{n=0}^{\mathcal{A}} a_n = v_0(R) = R(0) = \lim_{x \to 1} \sum_{n=0}^{+\infty} a_n x^n$$

Note that in this example we have for f, g in E

$$D(f.g) = (Df).g + v_0(f)Dg$$

then by induction we get for $n \geq 1$

$$D^n(f.g) = (D^n f).g + \sum_{k=0}^{n-1} v_0(D^k f)D^{n-k}g$$

Thus the convolution product of the sequences (a_n) and (b_n) is given by

$$(a \star b)_n = v_0(D^n(f.g)) = v_0(D^n f).v_0(g) + \sum_{k=0}^{n-1} v_0(D^k f)v_0(D^{n-k}g) = \sum_{k=0}^{n} a_k b_{n-k}$$

that is the usual Cauchy product of sequences.

Appendix A
Euler-MacLaurin and Euler-Boole Formulas

A.1   A Taylor Formula

The classical Taylor formula

$$f(x) = \sum_{k=0}^{m} \partial^k f(0) \frac{x^k}{k!} + \int_0^x \frac{(x-t)^m}{m!} \partial^{m+1} f(t) dt$$

can be generalized if we replace the polynomial $\frac{x^k}{k!}$ by other polynomials (Viskov 1988; Bourbaki 1959).

Definition If μ is a linear form on $C^0(\mathbb{R})$ such that $\mu(1) = 1$, we define the polynomials (P_n) by:

$$P_0 = 1$$
$$\partial P_n = P_{n-1} \, , \mu(P_n) = 0 \text{ for } n \geq 1$$

The Generating Function $\sum_{k \geq 0} P_k(x) z^k$

We have formally

$$\partial_x (\sum_{k \geq 0} P_k(x) z^k) = \sum_{k \geq 1} P_{k-1}(x) z^k = z(\sum_{k \geq 0} P_k(x) z^k)$$

© Springer International Publishing AG 2017
B. Candelpergher, *Ramanujan Summation of Divergent Series*,
Lecture Notes in Mathematics 2185, DOI 10.1007/978-3-319-63630-6

thus

$$\sum_{k \geq 0} P_k(x) z^k = C(z) e^{xz}$$

To evaluate $C(z)$ we use the notation μ_x for μ and by definition of (P_n) we can write

$$\mu_x \left(\sum_{k \geq 0} P_k(x) z^k \right) = \sum_{k \geq 0} \mu_x(P_k(x)) z^k = 1$$

$$\mu_x \left(\sum_{k \geq 0} P_k(x) z^k \right) = \mu_x(C(z) e^{xz}) = C(z) \mu_x(e^{xz})$$

this gives $C(z) = \frac{1}{\mu_x(e^{xz})}$. Thus the generating function of the sequence (P_n) is

$$\sum_n P_n(x) z^n = e^{xz} / M_\mu(z)$$

where the function M_μ is defined by $M_\mu(z) = \mu_x(e^{xz})$.

Examples

(1) $\mu(f) = f(0)$, $P_n(x) = \frac{x^n}{n!}$, $M_\mu(z) = 1$, $\sum_{n \geq 0} P_n(x) z^n = e^{xz}$

(2) $\mu(f) = \int_0^1 f(t) dt$, $P_n(x) = \frac{B_n(x)}{n!}$, $M_\mu(z) = \int_0^1 e^{zx} dx = \frac{1}{z}(e^z - 1)$

$$\sum_{n \geq 0} \frac{B_n(x)}{n!} z^n = \frac{z e^{xz}}{e^z - 1}$$

The $B_n(x)$ are the Bernoulli polynomials and the $B_n = B_n(0)$ the Bernoulli numbers. With the generating function we verify that $B_0 = 1$, $B_1 = -1/2$, $B_{2n+1} = 0$ if $n \geq 1$, $B_n(1-x) = (-1)^n B_n(x)$.

(3) $\mu(f) = \frac{1}{2}(f(0) + f(1))$, $P_n(x) = \frac{E_n(x)}{n!}$

$$\sum_{n \geq 0} \frac{E_n(x)}{n!} z^n = \frac{2 e^{xz}}{e^z + 1}$$

The $E_n(x)$ are the Euler polynomials and we set $E_n = E_n(0)$.

With the generating function we verify that $E_0 = 1$, $E_1 = -1/2$ if $n \geq 1$, $E_n(1-x) = (-1)^n E_n(x)$.

The Taylor Formula

Let f be a function in $C^\infty(\mathbb{R})$, then we have

$$f(x) = f(y) + \int_y^x \partial P_1(x+y-t)\partial f(t)dt$$

and by integration by parts we get for every $m \geq 1$

$$f(x) = f(y) + \sum_{k=1}^m (P_k(x)\partial^k f(y) - P_k(y)\partial^k f(x))$$

$$+ \int_y^x P_m(x+y-t)\partial^{m+1} f(t)dt$$

Applying μ to this function as a function of y gives a **general Taylor formula**: for every $m \geq 0$

$$f(x) = \sum_{k=0}^m \mu_y(\partial^k f(y))P_k(x) + \mu_y\left(\int_y^x P_m(x+y-t)\partial^{m+1} f(t)dt\right)$$

A.2   The Euler-MacLaurin Formula

We can transform the Taylor formula to get a summation formula. Taking $x = 0$ we get

$$f(0) = \sum_{k=0}^m \mu_y(\partial^k f(y))P_k(0) - \mu_y\left(\int_0^y P_m(y-t)\partial^{m+1} f(t)dt\right)$$

In the case of $\mu : f \mapsto \int_0^1 f(t)dt$ we have

$$f(0) = \sum_{k=0}^m \frac{B_k}{k!}\partial^{k-1} f]_0^1 - \int_0^1 \left(\int_0^y \frac{B_m(y-t)}{m!}\partial^{m+1} f(t)dt\right)dy$$

Replacing m by $2m$ and with $B_1 = -1/2$ and $B_{2k+1} = 0$, we get

$$f(0) = \int_0^1 f(t)dt + \frac{1}{2}(f(0) - f(1)) + \sum_{k=1}^m \frac{B_{2k}}{(2k)!}\partial^{2k-1} f]_0^1$$

$$- \int_0^1 \left(\int_0^y \frac{B_{2m}(y-t)}{(2m)!}\partial^{2m+1} f(t)dt\right)dy$$

The last integral can easily be evaluated by Fubini's theorem, we get

$$f(0) = \int_0^1 f(t)dt + \frac{1}{2}(f(0) - f(1)) + \sum_{k=1}^{m} \frac{B_{2k}}{(2k)!} \partial^{2k-1} f\big|_0^1$$
$$+ \int_0^1 \frac{B_{2m+1}(t)}{(2m+1)!} \partial^{2m+1} f(t)dt$$

Let j be a positive integer, by replacing f by $x \mapsto f(j+x)$ in the last formula, we have

$$f(j) = \int_j^{j+1} f(t)dt + \frac{1}{2}(f(j) - f(j+1)) + \sum_{k=1}^{m} \frac{B_{2k}}{(2k)!} \partial^{2k-1} f\big|_j^{j+1}$$
$$+ \int_j^{j+1} \frac{b_{2m+1}(t)}{(2m+1)!} \partial^{2m+1} f(t)dt$$

where $b_{2m+1}(t) = B_{2m+1}(t - [t])$.

Summing these relations for j from 1 to $n-1$, we get for $f \in C^\infty(]0, \infty[)$ the **Euler-MacLaurin formula**

$$f(1) + \ldots + f(n) = \int_1^n f(x)dx + \frac{f(1) + f(n)}{2}$$
$$+ \sum_{k=1}^{m} \frac{B_{2k}}{(2k)!} [\partial^{2k-1} f]_1^n$$
$$+ \int_1^n \frac{b_{2m+1}(x)}{(2m+1)!} \partial^{2m+1} f(x)dx$$

A.3   The Euler-Boole Formula

In the case of the Euler polynomials, the formula

$$f(0) = \sum_{k=0}^{m} \mu_y(\partial^k f(y)) P_k(0) - \mu_y\left(\int_0^y P_m(y-t) \partial^{m+1} f(t)dt\right)$$

gives

$$f(0) = \sum_{k=0}^{m} \frac{1}{2}(\partial^k f(0) + \partial^k f(1)) \frac{E_k}{k!} - \frac{1}{2} \int_0^1 \frac{(-1)^m E_m(t)}{m!} \partial^{m+1} f(t)dt$$

Let j be a positive integer, by replacing f by $x \mapsto f(j + x)$ in the last formula we obtain

$$f(j) = \sum_{k=0}^{m} \frac{1}{2}(\partial^k f(j) + \partial^k f(j+1))\frac{E_k}{k!}$$

$$-\frac{1}{2}\int_{j}^{j+1} \frac{(-1)^m E_m}{m!}(t-j)\, \partial^{m+1} f(t)\, dt$$

Let's define

$$e_m(t) = (-1)^{[t]}(-1)^m E_m (t - [t])$$

we obtain by summation on j the **Euler-Boole summation formula**

$$f(1) - f(2) + \ldots + (-1)^{n-1} f(n) = \frac{1}{2}\sum_{k=0}^{m} \partial^k f(1)\frac{E_k}{k!}$$

$$+\frac{(-1)^{n-1}}{2}\sum_{k=0}^{m} \partial^k f(n+1)\frac{E_k}{k!}$$

$$+\frac{1}{2}\int_{1}^{n+1} \frac{1}{m!}e_m(t)\, \partial^{m+1} f(t)\, dt$$

Appendix B
Ramanujan's Interpolation Formula and Carlson's Theorem

We give a proof of the following theorem.

Carlson's Theorem *Let f be an analytic function in the half-plane $Re(z) > -d$ where $0 < d < 1$. Let's assume there exist $a > 0$ and $b < \pi$ such that*

$$|f(z)| \le ae^{b|z|} \text{ for every } z \text{ with } Re(z) > -d$$

Then the condition $f(n) = 0$ for $n = 0, 1, 2, \ldots$, implies $f = 0$.

We prove this theorem by the use of an interpolation formula which is related to Ramanujan's interpolation formula.

First in Theorem 1 below we get an integral formula for the function

$$g : x \mapsto \sum_{n=0}^{+\infty} f(n)(-1)^n x^n$$

which is

$$g(x) = \frac{1}{2i\pi} \int_{c-i\infty}^{c+i\infty} \frac{\pi}{\sin(\pi u)} f(-u) x^{-u} du$$

Then in Theorem 2 we prove the interpolation formula

$$f(z) = \frac{-\sin(\pi z)}{\pi} \int_{0}^{+\infty} g(x) x^{-z-1} dx$$

Since for the definition of the function g we only need to know the values $f(n)$, $n = 0, 1, 2, \ldots$, we see that this interpolation formula determines the function f in the half-plane $Re(z) < -d$ when we only know the values $f(0), f(1), f(2), \ldots$, thus we have a proof of Carlson's theorem.

© Springer International Publishing AG 2017
B. Candelpergher, *Ramanujan Summation of Divergent Series*,
Lecture Notes in Mathematics 2185, DOI 10.1007/978-3-319-63630-6

Theorem 1 *Let f be an analytic function in the half-plane $Re(z) > -d$ where $0 < d < 1$. Let's assume there exist $a > 0$ and $b < \pi$ such that*

$$|f(z)| \le ae^{b|z|} \text{ for every } z \text{ with } Re(z) > -d$$

Then the series $\sum_{n\ge0} f(n)(-1)^n x^n$ is convergent in

$$D(0, e^{-b}) = \{x \in \mathbb{C}, |x| < e^{-b}\}$$

and defines an analytic function g in $D(0, e^{-b})$.
This function g has an analytic continuation in $S_b = \{|Arg(z)| < \pi - b\}$ which is defined by

$$g : z \mapsto \frac{1}{2i\pi} \int_{c-i\infty}^{c+i\infty} \frac{\pi}{\sin(\pi u)} f(-u)z^{-u} du$$

Proof For $0 \le \alpha < e^{-b}$ we have

$$|f(n)(-1)^n x^n| \le ae^{bn}\alpha^n$$

Thus the series $\sum_{n\ge0} f(n)(-1)^n x^n$ is normally convergent in $D(0, \alpha)$ and defines a function $g : z \mapsto \sum_{n=0}^{+\infty} f(n)(-1)^n z^n$ that is analytic in $D(0, e^{-b})$.

For $0 < x < e^{-b}$ and $0 < c < d$, the function $u \mapsto f(-u)x^{-u}$ is analytic in the half-plane $Re(u) < d$, and we consider the integral

$$\frac{1}{2i\pi} \int_{\gamma_N} \frac{\pi}{\sin(\pi u)} f(-u)x^{-u} du$$

where γ_N is the path

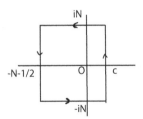

The function $u \mapsto \frac{\pi}{\sin(\pi u)} f(-u)x^{-u}$ has simple poles at $0, -1, -2, \ldots, -n, \ldots$ with

$$Res(\frac{\pi}{\sin(\pi u)}; -n) = (-1)^n f(n)x^n$$

thus we get

$$\frac{1}{2i\pi}\int_{\gamma_N}\frac{\pi}{\sin(\pi u)}f(-u)x^{-u}du = \sum_{n=0}^{N}f(n)(-1)^n x^n$$

Lemma 1 *When $N \to +\infty$ we have for $0 < x < e^{-b}$*

$$\frac{1}{2i\pi}\int_{\gamma_N}\frac{\pi}{\sin(\pi u)}f(-u)x^{-u}du \to \frac{1}{2i\pi}\int_{c-i\infty}^{c+i\infty}\frac{\pi}{\sin(\pi u)}f(-u)x^{-u}du$$

By the preceding lemma we get for $0 < x < e^{-b}$

$$\lim_{N\to+\infty}\frac{1}{2i\pi}\int_{\gamma_N}\frac{\pi}{\sin(\pi u)}f(-u)x^{-u}du = \lim_{N\to+\infty}\frac{1}{2i\pi}\int_{c-iN}^{c+iN}\frac{\pi}{\sin(\pi u)}f(-u)x^{-u}du$$

Thus we have for $0 < x < e^{-b}$

$$\sum_{n=0}^{+\infty}f(n)(-1)^n x^n = \frac{1}{2i\pi}\int_{c-i\infty}^{c+i\infty}\frac{\pi}{\sin(\pi u)}f(-u)x^{-u}du$$

Lemma 2 *For $0 < c < d$, the function*

$$g : \mapsto \int_{c-i\infty}^{c+i\infty}\frac{\pi}{\sin(\pi u)}f(-u)z^{-u}du$$

is defined and analytic in $S_b = \{|Arg(z)| < \pi - b\}$.
The function

$$z \mapsto \sum_{n=0}^{+\infty}f(n)(-1)^n z^n$$

is defined and analytic in $D(0, e^{-b}) = \{|z| < e^{-b}\}$ and is equal to g in the interval $[0, e^{-b}]$. By analytic continuation we get

$$\sum_{n=0}^{+\infty}f(n)(-1)^n z^n = \frac{1}{2i\pi}\int_{c-i\infty}^{c+i\infty}\frac{\pi}{\sin(\pi u)}f(-u)z^{-u}du \text{ if } z \in D(0, e^{-b}) \cap S_b.$$

□

The Mellin Inversion We now show that the formula

$$g(z) = \frac{1}{2i\pi}\int_{c-i\infty}^{c+i\infty}\frac{\pi}{\sin(\pi u)}f(-u)z^{-u}du$$

can be inverted to give

$$\frac{\pi}{\sin(\pi z)} f(-z) = \int_0^{+\infty} g(x) x^{z-1} dx$$

Theorem 2 *Let f be an analytic function in the half-plane $Re(z) > -d$ where $0 < d < 1$. Let's assume there exist $a > 0$ and $b < \pi$ such that*

$$|f(z)| \le a e^{b|z|} \text{ for every } z \text{ with } Re(z) > -d$$

For $0 < Re(z) < d$ we get

$$\frac{\pi}{\sin(\pi z)} f(-z) = \int_0^{+\infty} g(x) x^{z-1} dx$$

where g is the analytic continuation of the function $z \to \sum_{n=0}^{+\infty} f(n)(-1)^n z^n$ in $S_b = \{|Arg(z)| < \pi - b\}$.

We have in the half-plane $Re(z) > -d$ the interpolation formula

$$f(z) = \frac{-\sin(\pi z)}{\pi} \int_0^{+\infty} g(x) x^{-z-1} dx$$

Proof We consider $0 < c_1 < c_2 < d$, and z such that $c_1 < Re(z) < c_2$.

(a) First we evaluate $\int_0^1 g(x) x^{z-1} dx$.

We have

$$\int_0^1 g(x) x^{z-1} dx = \int_0^1 \left(\frac{1}{2i\pi} \int_{c_1-i\infty}^{c_1+i\infty} \frac{\pi}{\sin(\pi u)} f(-u) x^{-u} du\right) x^{z-1} dx$$

$$= \int_0^1 \left(\frac{1}{2\pi} \int_{-\infty}^{+\infty} f(-c_1 - it) x^{-c_1-it} \frac{\pi}{\sin(\pi(c_1 + it))} dt\right) x^{z-1} dx$$

Since

$$\left| f(-c_1 - it) \frac{\pi x^{-c_1-it} x^{z-1}}{\sin(\pi(c_1 + it))} \right| \le 2\pi a e^{bc} x^{-c_1} x^{Re(z)-1} e^{b|t|} \left| \frac{1}{e^{i\pi c_1} e^{-\pi t} - e^{-i\pi c_1} e^{\pi t}} \right|$$

with $Re(z) - c_1 - 1 > -1$ we get the integrability for $(t, x) \in \mathbb{R} \times [0, 1]$.

Thus by Fubini's theorem we get

$$\int_0^1 g(x) x^{z-1} dx = \frac{1}{2i\pi} \int_{c_1-i\infty}^{c_1+i\infty} \frac{\pi}{\sin(\pi u)} f(-u) \left(\int_0^1 x^{-u+z-1} dx\right) du$$

$$= \frac{1}{2i\pi} \int_{c_1-i\infty}^{c_1+i\infty} \frac{\pi}{\sin(\pi u)} f(-u) \frac{-1}{u - z} du$$

(b) Then we evaluate $\int_1^{+\infty} g(x)x^{z-1}dx$.

This is

$$\int_1^{+\infty} g(x)x^{z-1}dx = \int_1^{+\infty} \left(\frac{1}{2i\pi}\int_{c_2-i\infty}^{c_2+i\infty} \frac{\pi}{\sin(\pi u)}f(-u)x^{-u}du\right)x^{z-1}dx$$

As in (a) we see that for $\mathrm{Re}(z) - c_2 - 1 < -1$ we can apply Fubini's theorem to get

$$\int_1^{+\infty} g(x)x^{z-1}dx = \frac{1}{2i\pi}\int_{c_2-i\infty}^{c_2+i\infty} \frac{\pi}{\sin(\pi u)}f(-u)\left(\int_1^{+\infty} x^{-u+z-1}dx\right)du$$

$$= \frac{1}{2i\pi}\int_{c_2-i\infty}^{c_2+i\infty} \frac{\pi}{\sin(\pi u)}f(-u)\frac{1}{u-z}du$$

Finally we have for $c_1 < \mathrm{Re}(z) < c_2$

$$\int_0^{+\infty} g(x)x^{z-1}dx = \frac{1}{2i\pi}\int_{c_2-i\infty}^{c_2+i\infty} \frac{\pi}{\sin(\pi u)}f(-u)\frac{1}{u-z}du$$

$$- \frac{1}{2i\pi}\int_{c_1-i\infty}^{c_1+i\infty} \frac{\pi}{\sin(\pi u)}f(-u)\frac{1}{u-z}du$$

We then apply Cauchy's formula with the path

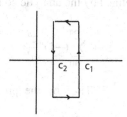

to get

$$\int_0^{+\infty} g(x)x^{z-1}dx = \frac{\pi}{\sin(\pi z)}f(-z)$$

By the preceding result we have for $-d < \mathrm{Re}(z) < 0$

$$f(z) = \frac{-\sin(\pi z)}{\pi}\int_0^{+\infty} g(x)x^{-z-1}dx$$

where g is the analytic continuation of the function $z \to \sum_{n=0}^{+\infty} f(n)(-1)^n z^n$.

The function

$$z \mapsto \frac{-\sin(\pi z)}{\pi} \int_0^{+\infty} g(x) x^{-z-1} dx$$

is defined and analytic in the half-plane $\mathrm{Re}(z) > -d$ since by the integral formula for $g(x)$ $(x > 0)$ we can write

$$\left| f(-c - it) x^{-c-it} \frac{\pi}{\sin(\pi(c + it))} \right| \leq 2\pi a e^{bc} x^{-c} e^{b|t|} \left| \frac{1}{e^{i\pi c} e^{-\pi t} - e^{-i\pi} e^{\pi t}} \right|$$

to get $g(x) = O(x^{-c})$ for $x \to +\infty$, for $0 < c < d$.

Thus by analytic continuation we get the *interpolation formula* in the half-plane $\mathrm{Re}(z) > -d$

$$f(z) = \frac{-\sin(\pi z)}{\pi} \int_0^{+\infty} g(x) x^{-z-1} dx$$

□

Remark: Ramanujan's Interpolation Formula

Let's consider the function $f : x \mapsto \frac{1}{\Gamma(x+1)}$. This function is analytic in the half-plane $Re(z) > -1$. The function g defined by the analytic continuation of

$$z \to \sum_{n=0}^{+\infty} (-1)^n \frac{z^n}{n!}$$

is simply the function $z \mapsto e^{-z}$. Thus by the preceding theorem we get for $0 < \mathrm{Re}(z) < 1$

$$\frac{\pi}{\sin(\pi z)} \frac{1}{\Gamma(1 - z)} = \int_0^{+\infty} e^{-x} x^{z-1} dx = \Gamma(z)$$

With $\Gamma(z)\Gamma(1 - z) = \frac{\pi}{\sin(\pi z)}$ we get for $0 < \mathrm{Re}(z) < d$

$$\int_0^{+\infty} g(x) x^{z-1} dx = \Gamma(z)\Gamma(1 - z) f(-z)$$

Let's take $h(z) = f(z)\Gamma(z+1)$. We have

$$g(x) = \sum_{n=0}^{+\infty} h(n)(-1)^n \frac{x^n}{n!}$$

and we get *Ramanujan's interpolation formula*

$$h(-z) = \frac{1}{\Gamma(z)} \int_0^{+\infty} g(x)x^{z-1}dx$$

Proofs of the Lemmas

Lemma 1 *When $N \to +\infty$ we have for $0 < x < e^{-b}$*

$$\frac{1}{2i\pi}\int_{\gamma_N}\frac{\pi}{\sin(\pi u)}f(-u)x^{-u}du \to \frac{1}{2i\pi}\int_{c-i\infty}^{c+i\infty}\frac{\pi}{\sin(\pi u)}f(-u)x^{-u}du$$

Proof (a) The integral on the vertical line through $-N - \frac{1}{2}$ is

$$\int_{-N}^{N} f(N+\frac{1}{2}-it)x^{N+\frac{1}{2}-it}\frac{\pi}{\sin(\pi(-N-\frac{1}{2}+it))}idt$$

since

$$\frac{\pi}{\sin(\pi(-N-\frac{1}{2}+it))} = \frac{2\pi(-1)^{N+1}}{e^{\pi t}+e^{-\pi t}}$$

we get

$$\left|\int_{-N}^{N} f(N+\frac{1}{2}-it)\frac{\pi x^{N+\frac{1}{2}-it}}{\sin(\pi(-N-\frac{1}{2}+it))}idt\right| \le 2\pi x^{N+\frac{1}{2}}\int_{-N}^{N}\frac{ae^{b(N+\frac{1}{2})}e^{b|t|}}{e^{\pi t}+e^{-\pi t}}dt$$

Since $b < \pi$ we have

$$\int_{-N}^{N}\frac{e^{b|t|}}{e^{\pi t}+e^{-\pi t}}dt \to \int_{-\infty}^{+\infty}\frac{e^{b|t|}}{e^{\pi t}+e^{-\pi t}}dt$$

For $0 < x < e^{-b}$ we have $b + Log(x) < 0$ thus

$$e^{b(N+\frac{1}{2})}x^{N+\frac{1}{2}} = e^{(N+\frac{1}{2})(b+Log(x))} \to 0$$

Thus the integral on the vertical line through $-N - \frac{1}{2}$ tends to 0 when $N \to +\infty$.

(b) The integral on the horizontal segment from $c + iN$ to $-N - \frac{1}{2} + iN$ is

$$I_N = -\int_{-N-\frac{1}{2}}^{c} f(-t - iN)x^{-t-iN}\frac{2i\pi}{e^{-\pi N}e^{i\pi t} - e^{\pi N}e^{-i\pi t}}dt$$

and we have

$$\left| f(-t - iN)x^{-t-iN}\frac{2i\pi}{e^{-\pi N}e^{i\pi t} - e^{\pi N}e^{-i\pi t}}\right| \le ae^{b(t+N)}x^{-t}\frac{2\pi}{e^{\pi N} - e^{-\pi N}}$$

thus

$$|I_N| \le ae^{bN}\frac{2\pi}{e^{\pi N} - e^{-\pi N}}\int_{-N-\frac{1}{2}}^{c} e^{t(b-Log(x))}dt$$

since $0 < x < e^{-b}$ we have $b - Log(x) > 2b > 0$ thus

$$\int_{-N-\frac{1}{2}}^{c} e^{t(b-Log(x))}dt \to \int_{-\infty}^{c} e^{t(b-Log(x))}dt$$

And $e^{bN}\frac{2\pi}{e^{\pi N}-e^{-\pi N}} \to 0$ since $b < \pi$.

(c) For the integral on the horizontal segment from $-N - \frac{1}{2} - iN$ to $c - iN$ the proof is similar to the preceding one. $\square$

Lemma 2 *Let's take $0 < c < d$, the function*

$$g : z \mapsto \int_{c-i\infty}^{c+i\infty} \frac{\pi}{\sin(\pi u)}f(-u)z^{-u}du$$

is defined and analytic in $S_b = \{|Arg(z)| < \pi - b\}$.

Proof For this integral on the vertical line through c we can write

$$\left| f(-c - it)\frac{\pi z^{-c-it}}{\sin(\pi(c + it))}i\right| \le 2\pi ae^{bc}|z|^{-c}e^{b|t|}e^{tArg(z)}\left|\frac{1}{e^{i\pi c}e^{-\pi t} - e^{-i\pi c}e^{\pi t}}\right|$$

For z in any compact K of S_b we have

$$e^{b|t|}e^{tArg(z)}\left|\frac{1}{e^{i\pi c}e^{-\pi t} - e^{-i\pi c}e^{\pi t}}\right| \le k(t)$$

where $t \mapsto k(t)$ is an integrable function independent of $z \in K$ since $|Arg(z)| < \pi - b'$ with $b < b' < \pi$.

The function

$$z \mapsto f(-c - it)z^{-c-it}\frac{\pi}{\sin(\pi(c + it))}i$$

is analytic for all t, thus we get the analyticity of the function defined by the integral. $\square$

Bibliography

T.M. Apostol, T.H. Vu, Dirichlet series related to the Riemann zeta function. J. Number Theory **19**, 85–102 (1984)

R. Ayoub, Euler and the Zeta function. Am. Math. Mon. **81**, 1067–1086 (1974)

R. Bellman, *A Brief Introduction to Theta Functions* (Holt, New York, 1961)

B.C. Berndt, *Ramanujan's Notebooks I, II, III, IV, V* (Springer, New York, 1985/1989/1991/1994/1998)

D. Birmingham, S. Sen, A Mellin transform summation technique. J. Phys. A Math. Gen. **20**, 4557–4560 (1987)

G. Boole, *A Treatise on the Calculus of Finite Differences*, 2nd edn. (Dover, New York, 1960)

J.M. Borwein, P.B. Borwein, K. Dilcher, Pi, Euler numbers and asymptotic expansions. Am. Math. Mon. **96**, 681–687 (1989)

J.M. Borwein, D.M. Bradley, R. Crandall, Computational strategies for Riemann zeta function. J. Comput. Appl. Math. **121**, 247–296 (2000)

N. Bourbaki, *Algebre Chapitre VI Développements Tayloriens généralisés. Formule sommatoire d'Euler-MacLaurin Hermann* (Hermann, Paris, 1959)

K.N. Boyadzhiev, H.G. Gadiyar, R. Padma, Alternating Euler sums at the negative integers and a relation to a certain convolution of Bernoulli numbers. Bull. Korean. Math. Soc. **45**(2), 277–283 (2008). arxiv.org/pdf/0811.4437

P.L. Butzer, P.J.S.G Ferreira, G. Schmeisser, R.L. Stens, The summation formulae of Euler-MacLaurin, Abel-Plana, Poisson and their interconnections with the approximate sampling formula of signal analysis. Res. Maths **59**, 359–400 (2011)

B. Candelpergher, Développements de Taylor et sommation des series. Exp. Math. **13**, 163–222 (1995)

B. Candelpergher, M.A. Coppo, E. Delabaere, La sommation de Ramanujan. L'Enseignement Mathematique **43**, 93–132 (1997)

B. Candelpergher, H.G. Gadiyar, R. Padma, Ramanujan summation and the exponential generating function $\sum_{k=0}^{\infty} \frac{z^k}{k} \zeta'(k)$. Ramanujan J. **21**, 99–122 (2010)

P. Cartier, *Mathemagics (A Tribute to L. Euler and R. Feynman)*. Lecture Notes in Physics, vol. 550 (Spinger, New York, 2000)

M.A. Coppo, Sur les sommes d'Euler divergentes. Exp. Math. **18**, 297–308 (2000)

M.A. Coppo, P.T. Young, On shifted Mascheroni series and hyperharmonic numbers. J. Number Theory **169**, 1–440 (2016)

H.M. Edwards, *Riemann Zeta Function* (Dover, New York, 1974)

H.M. Edwards, *Riemann Zeta Function* (Dover, New York, 2001)

© Springer International Publishing AG 2017

B. Candelpergher, *Ramanujan Summation of Divergent Series*,

Lecture Notes in Mathematics 2185, DOI 10.1007/978-3-319-63630-6

P. Flajolet, B. Salvy, Euler sums and contour integral representations. Exp. Math. **7**(1), 15–35 (1998)

E. Freitag, R. Busam, *Complex Analysis* (Springer, New York, 2009)

O. Furdui, *Limits, Series, and Fractional Part Integrals* (Springer, New York, 2012)

E. Grosswald, Comments on some formulae of Ramanujan. Acta Arith. **XXI**, 25–34 (1972)

G.H. Hardy, *Divergent Series* (Clarendon, Oxford, 1949)

D.H. Lehmer, Euler constants for arithmetical progressions. Acta Arih. **XXVII**, 125–142 (1975)

B. Malgrange, Sommation des séries divergentes. Expo. Math. **13**, 163–222 (1995)

N.E. Nörlund, Sur les séries d'interpolation (Gauthier-Villars, Paris, 1926)

S. Ramanujan, *Collected Papers*, ed. by G.H. Hardy, P.V. Seshu Aiyar, B.M. Wilson (Cambridge University Press, Cambridge, 1927)

J.-P. Ramis, Séries divergentes et théories asymptotiques. Bull. Soc. Math. France **121**(Panoramas et Synthŕses), 74 (1993)

B. Randé, *Les Carnets Indiens de Srinivasa Ramanujan* (Cassini, Paris, 2002)

R.J. Singh, D.P. Verma, Some series involving Riemann zeta function. Yokohama Math. J. **31**, 1–4 (1983)

R. Sitaramachandrarao, A formula of S. Ramanujan. J. Number Theory **25**, 1–19 (1987)

J. Sondow, Double integrals for Euler's constant and $ln(\frac{4}{\pi})$) and an analogue of Hadjicostas's formula. Am. Math. Mon. **112**(1), 61–65 (2005)

H.M. Srivastava, J. Choi, *Zeta and q-Zeta Functions an Associated Series and Integrals* (Elsevier, Amsterdam, 2012)

T. Tao, The Euler-MacLaurin formula, Bernoulli numbers, the zeta function, and real variable analytic continuation. terrytao.wordpress.com/2010/04/10/the-euler-maclaurin-formula-bernoulli-number (2010)

E.C. Titchmarsh, D.R. Heath-Brown, *The Theory of the Riemann Zeta-Function* (Clarendon, Oxford, 2007)

V.S. Varadarajan, Euler and his work on infinite series. Bull. Am. Math. Soc. **44**(4), 515 (2007)

I. Vardi, *Computational Recreations in Mathematica* (Addison-Wesley, Redwood City, CA, 1991)

O.V. Viskov, A non commutative approach to classical problems of analysis. Proc. Steklov Inst. Math. **4**, 21–32 (1988)

K. Yoshino, Difference equation in the space of holomorphic functions of exponential type and Ramanujan summation. ams.org.epr.uca/ma/mathscinet/search

D. Zagier, Valeurs des fonctions zéta des corps quadratiques réels aux entiers négatifs. Société Mathématique de France. Astérisque **41/42**, 133–151 (1977)

D. Zagier, Values of zeta functions and their applications, in *First European Congress of Mathematics Volume II*. Progress in Mathematics, vol. 120 (Birkhauser, Basel, 1994)

N.Y. Zhang, K. Williams, Some results on the generalized Stieltjes constants. Analysis **14**, 147–162 (1994)

Index

© Springer International Publishing AG 2017
B. Candelpergher, *Ramanujan Summation of Divergent Series*,
Lecture Notes in Mathematics 2185, DOI 10.1007/978-3-319-63630-6

LECTURE NOTES IN MATHEMATICS    Springer

Editors in Chief: J.-M. Morel, B. Teissier;

Editorial Policy

1. Lecture Notes aim to report new developments in all areas of mathematics and their applications – quickly, informally and at a high level. Mathematical texts analysing new developments in modelling and numerical simulation are welcome.

   Manuscripts should be reasonably self-contained and rounded off. Thus they may, and often will, present not only results of the author but also related work by other people. They may be based on specialised lecture courses. Furthermore, the manuscripts should provide sufficient motivation, examples and applications. This clearly distinguishes Lecture Notes from journal articles or technical reports which normally are very concise. Articles intended for a journal but too long to be accepted by most journals, usually do not have this "lecture notes" character. For similar reasons it is unusual for doctoral theses to be accepted for the Lecture Notes series, though habilitation theses may be appropriate.

2. Besides monographs, multi-author manuscripts resulting from SUMMER SCHOOLS or similar INTENSIVE COURSES are welcome, provided their objective was held to present an active mathematical topic to an audience at the beginning or intermediate graduate level (a list of participants should be provided).

   The resulting manuscript should not be just a collection of course notes, but should require advance planning and coordination among the main lecturers. The subject matter should dictate the structure of the book. This structure should be motivated and explained in a scientific introduction, and the notation, references, index and formulation of results should be, if possible, unified by the editors. Each contribution should have an abstract and an introduction referring to the other contributions. In other words, more preparatory work must go into a multi-authored volume than simply assembling a disparate collection of papers, communicated at the event.

3. Manuscripts should be submitted either online at www.editorialmanager.com/lnm to Springer's mathematics editorial in Heidelberg, or electronically to one of the series editors. Authors should be aware that incomplete or insufficiently close-to-final manuscripts almost always result in longer refereeing times and nevertheless unclear referees' recommendations, making further refereeing of a final draft necessary. The strict minimum amount of material that will be considered should include a detailed outline describing the planned contents of each chapter, a bibliography and several sample chapters. Parallel submission of a manuscript to another publisher while under consideration for LNM is not acceptable and can lead to rejection.

4. In general, **monographs** will be sent out to at least 2 external referees for evaluation.

   A final decision to publish can be made only on the basis of the complete manuscript, however a refereeing process leading to a preliminary decision can be based on a pre-final or incomplete manuscript.

   Volume Editors of **multi-author works** are expected to arrange for the refereeing, to the usual scientific standards, of the individual contributions. If the resulting reports can be forwarded to the LNM Editorial Board, this is very helpful. If no reports are forwarded or if other questions remain unclear in respect of homogeneity etc, the series editors may wish to consult external referees for an overall evaluation of the volume.

5. Manuscripts should in general be submitted in English. Final manuscripts should contain at least 100 pages of mathematical text and should always include

   – a table of contents;
   – an informative introduction, with adequate motivation and perhaps some historical remarks: it should be accessible to a reader not intimately familiar with the topic treated;
   – a subject index: as a rule this is genuinely helpful for the reader.
   – For evaluation purposes, manuscripts should be submitted as pdf files.

6. Careful preparation of the manuscripts will help keep production time short besides ensuring satisfactory appearance of the finished book in print and online. After acceptance of the manuscript authors will be asked to prepare the final LaTeX source files (see LaTeX templates online: https://www.springer.com/gb/authors-editors/book-authors-editors/manuscriptpreparation/5636) plus the corresponding pdf- or zipped ps-file. The LaTeX source files are essential for producing the full-text online version of the book, see http://link.springer.com/bookseries/304 for the existing online volumes of LNM). The technical production of a Lecture Notes volume takes approximately 12 weeks. Additional instructions, if necessary, are available on request from lnm@springer.com.

7. Authors receive a total of 30 free copies of their volume and free access to their book on SpringerLink, but no royalties. They are entitled to a discount of 33.3 % on the price of Springer books purchased for their personal use, if ordering directly from Springer.

8. Commitment to publish is made by a *Publishing Agreement*; contributing authors of multiauthor books are requested to sign a *Consent to Publish form*. Springer-Verlag registers the copyright for each volume. Authors are free to reuse material contained in their LNM volumes in later publications: a brief written (or e-mail) request for formal permission is sufficient.

Addresses:
Professor Jean-Michel Morel, CMLA, École Normale Supérieure de Cachan, France
E-mail: moreljeanmichel@gmail.com

Professor Bernard Teissier, Equipe Géométrie et Dynamique,
Institut de Mathématiques de Jussieu – Paris Rive Gauche, Paris, France
E-mail: bernard.teissier@imj-prg.fr

Springer: Ute McCrory, Mathematics, Heidelberg, Germany,
E-mail: lnm@springer.com

Printed in the United States
By Bookmasters